ROCKETS MISSILES UFOs

AND OTHER INTERESTING TOPICS

By: Prof. Juan P. Rivera

INFINITY PUBLISHING

ISBN 978-0-7414-9861-8 Paperback
ISBN 978-0-7414-9862-5 eBook

Printed in the United States of America

Published October 2013

INFINITY PUBLISHING
1094 New DeHaven Street, Suite 100
West Conshohocken, PA 19428-2713
Toll-free (877) BUY BOOK
Local Phone (610) 941-9999
Fax (610) 941-9959
Info@buybooksontheweb.com
www.buybooksontheweb.com

TABLE OF CONTENTS

INDEX OF IMAGES

DEDICATION

I dedicate this book first to my wife Nora M. Rivera, who has accompanied me on most trips narrated in this book. I want to express my gratitude to my three children, Juan Carlos, Ricardo and the youngest, Eduardo Gabriel, who have been my source of inspiration to fight against all odds in order to bring home the daily bread. Finally, I dedicate this book to people that I call unsung heroes who have also been the source of inspiration to undertake a task of such magnitude.

NOTES TO THE READER

My life has had a particularly intriguing trajectory. Looking back, I have been extremely fortunate in my life from birth until the time I began writing this book. In this book, I narrate my experiences at the many places where I have been. It also deals with many issues that I hope will be entertaining to the reader.

The table of contents lists the topics that describe the critical incidents and events of my life. Each one will be discussed in more detail in this book. When the reader has finished my book, he will learn the secrets of my success. If the reader applies the same principles to his own life, he can succeed too.

Despite the poor and humble background of my birth, if one has faith in God first and in oneself second, one can overcome vices and the temptations of evil and become a person worthy of prestige in the community. Again, I hope this book will stimulate new generations to learn that once a target is chosen, with patience and faith in God, any obstacles can be overcome along the way.

My philosophy of life has always been: (1) to realize that your real parents are not the ones who brought you into this world, but rather the ones who raised you to adulthood. Raising a child or children is not easy. Not only does one have to provide shelter, food, and clothing. One also has to deal with unexpected situations such as diseases, accidents, problems in school, and so on; (2) to provide a defense system when children are attacked by other children or adults; (3) to teach children the fear of God or a system of moral values in order to respect themselves and others.

PROLOGUE

Life is a mystery that unfolds like the pages of a book. We know the time that we were born, but we do not know the day that we will die. As that date approaches, we look at the past, and our memories unfold like a tape on a video recorder. As I write this book, I feel many emotions. Some of my memories bring joy others bring tears.

My life has been extremely challenging, and I hope that this book will inspire future generations. In this book, I reveal the secrets that I used to be successful in life: Do it right the first time! That way I do not waste my time going back to do something that should have been done right in the first place. Faith in God and the study of Christianity have been incredible fortresses that have helped me overcome the inertia created by some of the enormous problems that I have encountered in my long life. The study of mathematics and physics taught me to learn methods to solve complex scientific and physics problems. Those same methods, a bit altered, I have used to solve many problems of my daily life. Whether they are business, health-related, or religious, I have used the same scientific reasoning to solve them. The first step is to understand the problem. If one does not, it is difficult to come to the correct solution.

Everything that I have done in my life has been done with a lot of work and tears. In this book, I reveal the customs, the music, the politics, the wars, the student protests that I saw in my youth. I reveal my meetings with prominent figures like rocket scientists. I explain the basic concepts of missiles and radar in laymen's terms. I recount my days working both as a scientist and an engineer in the Space Shuttle Program. I explain my experiences with the development of the International Space Station Program. Understanding the challenges that we faced in the development of space vehicles will help in the future development of more modern space vehicles. Maybe one day we will be able to return to the moon or commence deeper space exploration. I, most likely, will not live to see future developments, but many of the readers may. Finally, I describe my encounter with aliens and a UFO while I was employed as a scientist at White Sands, New Mexico. What were extraterrestrial beings doing one night in the middle of a desert? I have kept this buried in my heart for more than 30 years, but now I can tell the secret to my readers.

It is necessary to understand the role that Puerto Rico played in my childhood during and after the Second World War. During that time, the Puerto Rican way of life was remarkably different from what it is now. The discussion is rather detailed at times and, on other occasions, is presented in the form of a summary. Either way, I will demonstrate how I turned failure into success.

The important thing for me has always been to formulate a purpose then to work toward that goal. When I get up every morning, I make a few notes of what I want to do during the day. Before going to bed, I examine what I did that day and what I failed to do. Whatever I did not do today will be the first items on my list for the next day. Then I add new items to my list. As a boy, I learned not to leave for tomorrow what I can do today.

I am immensely grateful to my teachers who prepared me for admission to the University of Puerto Rico and also to those professors at Columbia University who pushed me to the limit. I had an extremely difficult time when I was living in New York City as a graduate student. I met the best teachers of physics and mathematics from whom I obtained an outstanding education. When I went to the National Autonomous University of Mexico in Mexico City, I met Mexican teachers of the highest caliber from whom I learned many things in the field of modern physics and nuclear physics. Then I went to the University of New Mexico in Albuquerque, New Mexico, where I learned more about astronomy and the physics of outer space. It was like obtaining another doctorate in another branch of physics. All that academic accumulation was additional stimulation for me to meet the enormous challenges at all the different places where I was employed during my professional life. As one continues reading this book, one will learn more details about the places and events that forged the person I am today.

Juan Pedro Rivera
Orlando, Florida
USA

Figure 1: Here the author is shown with the space shuttle Atlantis in the background before the first launch.

BREAD, LAND AND FREEDOM AND LUIS MUñOZ MARIN

I grew up in a neighborhood of extremely poor people in the city of Rio Piedras, Puerto Rico. I was born on January 25, 1938, in a modest home owned by my parents. My father was Pedro Rivera de Gracia, a carpenter by trade, and my mother was Benigna Diaz Hernández, a housewife most of her life. My parents had 11 children, seven boys and four girls. I was in the middle—five before and five after me.

In the 1930s and 40s, Rio Piedras was an autonomous municipality, as were the municipalities of other villages on the Island. At that time, Puerto Rico was governed by a military governor appointed by the President of the United States, Franklin D Roosevelt. There were some political parties that had their own leaders, and a political forum existed where the various leaders were able to express their political ideas. The Nationalist Party leader in Puerto Rico was Don Pedro Albizu Campos; the leader of the Republican Party was Don Miguel Angel Garcia Méndez, and the Independence Party leader was Gilberto Concepción de Gracia. The Popular Democratic Party leader was Don Luis Muñoz Marin. Previously there were other political parties, led by others, but they eventually faded away. My father spoke of the Socialist Party, the Unionist Party and others. In the political struggle of the early 1940s, the Popular Democratic Party forged ahead of the others, principally due to the charisma of its leader, Don Luis Muñoz Marin, with the slogan, BREAD, LAND AND FREEDOM.

At that time, the economic power was in the hands of large landowners, especially in the sugar cane industry. The landowners who planted tobacco were also highly influential in the power struggle. Above all, there were other North American companies that held the political clout in the United States Congress. These companies controlled the coffee industry, the tobacco industry and the sugar cane industry. The poor workers toiled from sunrise to sunset, for a miserable wage, and whatever they earned was owed to the landowner's shop. The Puerto Rican "jibaro" was exploited by large landowners; Don Luis Muñoz Marin promised the workers what they longed for. Muñoz's philosophy was based on the following ideas: (1) the land must be owned by those who work it; (2) the harvest should bring food to the workers' tables; and (3) Puerto Rico should be an independent, autonomous Island, free from the political masters in Washington. Those were the key ideas of Don Luis Muñoz Marin. He gave the people what other leaders proclaimed only within their own political parties. The Independence Party remained almost without a political slogan. In fact, Muñoz took the symbol of the flag of Puerto Rico, which was the same flag that the "independentistas" and nationalists were flying at their homes. A banner bearing the likeness of a Puerto Rican "jibaro", painted in red in the center of a white rectangular fabric, became the flag of the Popular Democratic Party of Muñoz Marin. The main difference between the Independence Party and the Nationalist Party was that the Nationalists wanted to bring freedom for Puerto Rico by force. I remember hearing Don Pedro Albizu Campos saying at his rallies that freedom is not brought by votes, but has to be won by any means. Few people knew that Don Pedro Albizu Campos was a colonel in the U.S. Army and was an Army veteran. Albizu was black, and it had been said that he was humiliated when he was in the army, despite having a high rank. Albizu was a graduate of Harvard and spoke seven languages. One of the best speakers of Puerto Rico, his oratory skills, was highly praised. Don Miguel Garcia Méndez was second in public speaking, and Don Gilberto Concepción de Gracia was third. Don Luis Muñoz Marin came in last because the others were considered better speakers according to the experts. When I started listening to these prominent political leaders of Puerto Rico, I remember that, although they were political enemies, they treated each other as Puerto Rican brothers. It is a lesson all political leaders after them would do well to follow. It is extremely discouraging to watch the news and see today's political leaders acting as if they were in the ghetto.

The Popular Democratic Party was highly esteemed during my childhood, and, under the leadership of Muñoz Marin, worked exceptionally hard to bring about the best for Puerto Rico. In 1948, U.S. President Harry S. Truman appointed Don Jesus T. Piñero as the first Puerto Rican governor becoming the first governor of Puerto Rican blood. Later on, the Popular Democratic Party won the elections, and the distinguished Don Luis Muñoz Marin became the first Puerto Rican governor elected by the people. Muñoz's salary was $12,000 per year. He was an unusually honest person, and those around him gained the respect of the majority of the Puerto Rican people. In 1952, Don Luis Muñoz Marin and his Popular Democratic Party created the Commonwealth of Puerto Rico.

The Jones Act of 1917 gave U.S. citizenship to Puerto Ricans. All persons of Puerto Rican blood or born after that year automatically acquired U.S. citizenship; however, there was a price to pay which will be explained later.

Doña Felicia Rincón de Gautier was the first mayor of the city of San Juan. "Doña Fela," as we called her, was unquestionably a shrewd politician. She decided to dissolve the city of Rio Piedras and annex it to the municipality of San Juan. There was an election where the mayor of Rio Piedras opposed Doña Fela. Mayor Alvarez lost that election, and the city of Rio Piedras was dissolved. As a child, I remember that there were many people living in Rio Piedras who did not want to be part of the municipality of San Juan. Don Ramon S. Torres was a business owner who ran his own radio show every Sunday defending his ideas. Unfortunately, it was like preaching in the desert.

As they say, every cloud has a silver lining, so the annexation of Rio Piedras into the municipality of San Juan turned out to be a tremendous gift. It brought rapid progress into the neighborhoods of Rio Piedras. All the streets were tarred, and lights were installed at all major intersections to control the flow of traffic. They kept the streets of Rio Piedras clean with garbage collection twice a week. Doña Fela brought electricity to all neighborhoods. She improved all public schools. Clinics were built; hospitals were improved and were free of charge for poor people. Those who possessed a "health card" could use it to pay for doctors or medicines.

The homes of most poor people displayed a Popular Democratic Party flag outside and a picture of Muñoz Marin in the living room or kitchen. This was soon imitated by the majority of people in Puerto Rico. At election time, Popular Democratic Party flags were displayed profusely in the poorest neighborhoods. The governor put in place ways to improve public transportation all over the Island. In short, Puerto Rico saw tremendous growth during the 1950s and 60s. Muñoz Marin created the Land Authority. The lands of large property-owners became the property of the Land Authority, which then divided the land into small plots. These plots, known as "parcelas," were given to "parceleros," people who could prove that they could grow food and build a house on the land. Muñoz Marin, through what became known as the Bank of Farmer Homes, lent money to each settler so that he could build a house on his land. At first, the plot was owned by the Land Authority, but years later, Don Luis Ferre, another governor of Puerto Rico, gave the title of ownership to the parceleros for one dollar. Don Luis A. Ferre was also one of the best governors of Puerto Rico. A wealthy man in his own right, he donated his entire salary to charity.

When Muñoz grew older, things started changing on the political scene, and the wealth of the Island began to decline. According to the press, some leaders surrounding Muñoz were involved in political scandals. Then began an exodus of Puerto Ricans leaving the Island for better opportunities in cities like New York, Chicago, and Los Angeles—an exodus that would be repeated several times in later years.

Many Puerto Ricans do not know the history of Puerto Rico; therefore, I think it is crucial for me to revisit some of the events in our history. Christopher Columbus discovered America on October 12, 1492. He landed on an island that he called San Salvador, which today, is part of the archipelago of the Bahamas. It is officially accepted that on November 19, 1492, Christopher Columbus and his crew arrived at a port he called San Juan Bautista. Today it is the capital of Puerto Rico.

As to the official landing site on the Island, there are several versions about where the ships anchored. Some historians say that the landing was in Aguadilla; others say it was in Guayanilla, Mayaguez, Anasco, Rincon, or Boqueron. The main thing is that Christopher Columbus was the first white man to set foot in Puerto Rico.

Greed took over the hearts of the settlers. The only things that mattered to them were gold, silver, other precious metals, diamonds and other gemstones. The Taino Indians were not used to being slaves. The Caribe Indians had a history of being a brave and warlike people who liked to fight and defend their honor. They were also cannibals. Not wanting to face the formidable Spaniards, the Caribe Indians decided to flee to other islands south of Puerto Rico. Many of them settled in other islands of the Caribbean; others settled along the Caribbean coast of South America.

The Spanish conquerors were forced to bring slaves from Africa. They came mostly from Nigeria and other neighboring countries. Later on came a wave of buccaneers and privateers from France, Holland, and England. The Spaniards were forced to build large forts to defend the Island of Puerto Rico and other islands that they had colonized. They also brought with them plagues of Venereal diseases that ravaged the lands they conquered. The Catholic Church turned a blind eye to the ravages inflicted by the colonizers—all in the name of the Church and the Spanish Empire. When gold and precious metals became scarce, the Spaniards were forced to colonize other lands, eventually moving on to Hispaniola, Cuba, and other islands of the Caribbean. In their pursuit of gold, they went to Central and South America.

When the Americans came to the Puerto Rican shores in 1898, the Island of Puerto Rico had no metals or gemstones. All that remained were the fields of sugar cane, tobacco, and other agricultural products. A new era began in Puerto Rico's history as a colony of the USA and continues right up to the present.

At the beginning of the 20th century, the economic landscape of Puerto Rico was extremely poor. Coffee and tobacco became the staples of the agrarian economy. Everything was controlled by the new landowners and large American companies. This same scenario was repeated in other Caribbean islands and Central and South American countries. Of course, the United States was not the only imperialist country. The governments of England, France, Italy, and Portugal had their own imperialist agendas. The list is long.

In 1898, the Island became a territory of the United States and remained so to this day. As a U.S. colony, Puerto Rican political status is dependent on the U.S. Congress. While the Island has enjoyed varying degrees of some local autonomy, there has been no change in "maritime trade laws," which means that the Island does not have a free market. We are a captive market. We get almost everything from the United States. There are some exceptions, but the majority of products come from the mainland. Also, there has been no change in the immigration laws. The Island has been invaded by people from neighboring islands to a greater extent and people of Central and South America to a lesser extent. This is certainly a serious problem, and no government has figured out how to fix this. We have had several plebiscites that ultimately do nothing more than act as sedatives, calming people's emotions. (The same situation occurs with Native American Indians. They stay on their own parcels of land, but the conditions of poverty and other problems persist, and nobody knows the answer.) It is estimated that from 1900 until around 1940, American corporations controlled agriculture. In 1952, when Muñoz created a kind of socialism, the large corporations of the United States moved to control other market segments or went elsewhere.

In the first part of the 20th century, the Island was hit by a series of hurricanes that ravaged agriculture. I remember how older people spoke of "San Ciriaco," "San Ciprian" (1928), "San Felipe" (1932), which were the names of those hurricanes. In those days, in Puerto Rico, there was a tradition to give children the name of the saint on whose day they were born. Storms were also given the name of the saint on whose day they hit the Island. At that time, 365 pages of the calendar were in every home and marked the passing of people and hurricanes.

The Great Depression in North America during the 1930s affected the economy of the Island. Many municipalities lacked the most basic services such as schools, hospitals, electricity, drinking water, health services, roads, and recreational facilities. The government did not have the resources needed to support those facilities. Probably they were just not a priority for them at that time. In the cane fields, the workers' living conditions were even worse. They lived as best as they could, from day to day, supporting one another or begging for help from the landowner, without any real hope that living conditions would be better any time in the future. Into this world of poverty, I was born and grew up.

POVERTY IN POOR AREAS OF PUERTO RICO

In the 1930s, both the U.S. and Puerto Rico suffered horrendous economic problems. Not only was there hunger in America, but there was a lot of hunger in Puerto Rico. Unfortunately, I came into the world when there were all these economic problems. Luckily, my father was a carpenter, and later he became a kind of civil engineer. My father went to school only up to the eighth grade. At that time, an eighth grade education was like having a high school diploma because teaching was a hodgepodge of the Spanish and American school systems. My father began to build houses, buildings, and even bridges. Although wood was cheaper, concrete was used, which was not customary in new buildings on the Island. At the beginning of the 1940s, he made a fortune and became one of the richest people in the city of Rio Piedras.

My godmother and godfather lived beside my father's property in a poor neighborhood of Rio Piedras. My godmother took care of me after my birth, and I had a lot of affection for her. Her husband, Lolo, who was my father's brother, was seriously ill because of a fall that possibly caused him a brain hemorrhage. In those days (the late 1930s), medical treatment was not readily available.

I've always had a marvelous memory, and though I was less than three years old at the time, I remember seeing my uncle Lolo walking with considerable difficulty, holding on to the walls in order to go to the latrine which was outside the house. One day I picked up a few herbal plants and made tea for my uncle. From his bed, surrounded by other people in the family, he told everyone, "take care of this baby because he is going to be something great when he grows up." Every day I witnessed my uncle's life fading (this memory still brings tears to my eyes). I still remember that day when my uncle Lolo passed away in 1941. I think he was only 39 when he died. I remember this date because shortly afterward Japan declared war on the United States.

My godmother, Josefa Colon Miranda, now a widow at an early age, went to my parents to ask them if she could raise me to replace the loneliness of losing my uncle. At that time, it was not unusual to let someone else raise a child without the necessity of legal papers. My parents agreed to my foster mother's request. In her late 30's at the time, she raised me as her own child. I was about three years old when this transfer of "ownership" occurred.

My parents, brothers, and sisters moved to an exclusive Rio Piedras neighborhood known as Santa Rita, where the rich people lived. I stayed behind in a poor neighborhood considered one of the slums of the big city. As I grew up, I would see my family sometimes. I was around five years old when Doña Pepa (my foster mother) told me one day, "Baby I'm going to tell you something extremely important. You are not my true son; you are my foster son. Your dad is Pedro, and your real mother is Benigna. Tell me if you want to go with them or stay with me." Hearing those words was extremely difficult for me, because I had to make a decision at an early age that would undoubtedly affect my future. I knew who Pedro and Benigna were. Not knowing what to say, I told Doña Pepa, "I'll decide tomorrow." That night I could not sleep. On one hand, my father was an immensely wealthy man, and Doña Pepa was a poor widow. On the other hand, I was the center of her life and my foster siblings, Toña, Carmelito, and Manuel loved me and were close to me. What should I do? If a made the wrong choice, my life might be ruined!

The next day I told Doña Pepa that I had made my decision. I said, "I know that Benigna and Pedro are my real parents because I come from their flesh; the blood in my veins is truly theirs. I want to stay with you but be able to see my parents and my blood brothers and sisters from time to time. Then she said. "Well, baby your will is my will, and I am an extremely poor woman, but I will provide you with all the necessities that you need. If I have to eat dirt, I will do that with pleasure, so that one day you will have a decent education. I want you to achieve the education that I never had. I barely know how to read and write, and I am a widow and poor but healthy and strong. You are the person that will be with me until I become an old lady. Later on you will abandon me for another woman, but I did my job as a good mother."

Doña Pepa always took me with her wherever she went because I was a child and she did not trust anyone to look after me. That was a tremendous plus for me because I got to see the whole family—her side and my mom and dad's side. My uncles and aunts always asked the same question: "Who do you love more, Pepita or Benigna?" But I was extremely astute. I always answered, "I love both of them the same."

I never lacked for clothes, shoes, or food. When I was sick, I always had the best doctors. I was fortunate enough to attend kindergarten through high school (1944 until 1956) in one of the best private schools at that time, Colegio San Antonio in Rio Piedras. The majority of my schoolmates came from the wealthy families of my hometown. My godmother, Doña Pepa, was immensely proud of me, and I brought joy to her life. Her sons and daughter were also proud of me because I was fulfilling their dreams. Due to living conditions at the time, they did not have the possibility that I had of pursuing a decent education. The only other one who was able to get a decent education was my oldest half-brother, Carmelito, who, later on, became a Certified Public Accountant. My half-sister Toña got married, and my other half-brother, Manuel, joined the Air Force and later became a draftsman by trade.

During my years at the private school, I was a guy who did not study. Everything I heard was recorded in my head. I rarely took notes in class; I never studied for a test. Everything came easily to me. I had a photographic memory and my mind was like a video player. I was able to play back in my mind the events of the day. I thought at the time that this was a natural thing for everybody. It was later in life I became aware that this skill was a unique gift. My inclination was towards mathematics and science. I loved algebra and both plane and solid geometry. I enjoyed learning about science. During high school, I enjoyed physics, biology, and chemistry. At that time, both mathematics and science were included as requirements to obtain a high school diploma.

I was a child living in a suburb of Rio Piedras, where even police dared not enter. The streets were not paved. A path full of mud and sludge formed when it rained so that people were unable to walk. There were several cafes and grocery stores where people would buy what they needed. Pedro Mangual and his wife Teresa owned a corner where the largest building in the neighborhood stood. They lived on the second floor of what became known as Pedro Mangual's Corner. The first floor on one side had a grocery store and on the other side a bar and a coffee shop.

On weekends, the neighborhood kids hung out drinking sugar cane rum and oversized bottles of beer. Once the alcohol had gone to their heads, fights began. They fought with machetes, knives, or revolvers. As there was no radio or television at that time, it became an event where crowds of people gathered to watch the exciting fights. The city hospital was at the end of the street where I lived. From my balcony, I watched all varieties of weekend flights. In the end, the losers were taken to the hospital, jail or the graveyard. My mom would not let me watch the fights, so I had to look for a way to escape from her. I used to see those fights as part of my daily life, and seeing blood run was extremely exciting for me. Now I understand how the Roman emperors must have enjoyed the bloody battles at the Coliseum in Rome. In my case, it was hard to look the other way because the roar of the people was so intense that I was attracted to those fights like a magnet.

At that time, there were no ambulances, or if there were, they were for rich people. So the wounded were carried either in hammocks or on makeshift stretchers, or Don Pedro Mangual, one of the few who had a car and a truck at that time, carried the wounded on the platform of his truck.

Facing Mangual's Corner, there was a barber shop whose owner was called Thomas. At that time, barber shops were the center of news and gossip, but barbers also provided other services like pulling a tooth or molar. There were many people who could not afford to pay a professional dentist. Fortunately, Thomas did an excellent job. He was a strong man who was able to hold a man's head, force his mouth wide open, and pull a tooth for a dollar. This particular hold became known as "Thomas's sleepy hold," which described the way to grab a person, so they were not able to move until the tooth was pulled. As no pain killers were used, Thomas would pour a little rum into the empty tooth socket to relieve the pain. There was another barber by the "Hill of the Wind," which was another part of the neighborhood. For a dime, he would cut hair but not comb it. A haircut with styling and fragrance cost a quarter. Children sat in the barber's lap, to get a haircut for a nickel. I refused to get a haircut from this guy because there were rumors that he was a child molester.

To heal stomach aches, colds, fever and so on, they brewed a tea concoction made from plant leaves. There was a woman called "the witch" who healed people by using herbal remedies. They cooked pigeons to make broth for postpartum women. For wounds and cuts, chromium-mercury or tincture of iodine were used. For an earache, they crushed a plant on a stone, and then the juice extracted was poured in the ear to remove ear wax. Then they blew in your ear. That was the specialty of another woman who was called Doña Yuya.

Doña Pepa, my godmother, was a seamstress. She owned one of those old Singer machines with the antiquated pedal on the bottom. Her skill with the machine brought food to the table. She taught her daughter Toña how to sew, and Toña became an expert seamstress in everything related to fabrics. She could replicate a dress by just looking at a picture in a Sears catalog. I loved watching them work on that old Singer.

In the neighborhood, there were a number of people who had proficiency in a variety of skills and services. There were shoemakers, seamstresses, barber-dentists, midwives—enough talent so that no one needed to go to the city. There were also all types of businesses: grocery stores, coffee shops, funeral homes, carpentry, car mechanics, and many more.

There were some guys who were called "malotes." They were people who did not fear death. They gathered on street corners watching people go by, choosing their victims. Mostly they asked for a nickel or a quarter or to buy them a shot of rum or a beer. Otherwise, they put a large knife called a "sevillana" to your throat. Their daily job was to hang out in that corner selecting their victims.

Others were engaged in stealing. There was one called "Cachipa." Cachipa was a "decent crook." After he stole a chicken, he invited you to his place to share it. People hung clothes to dry and then a guy who was called "Cachiche" stole them. There was a bicycle rental agency that rented a bike for twenty-five cents per hour. My cousin Pepe rented a bike for an hour, paying a quarter. Then he sold it to Reuben. When Reuben looked the other way, my cousin stole it back and sold it again. Pepe told me that he sold the bike five times making an exorbitant profit. Next day, he returned the bike to the rental agency in the middle of the night.

My godmother only knew how to write her name, so I had to read the mail for her. She had a terrible temper with other people, but always treated me with love and respect. She may have once had thoughts of re-marrying, but as they said in those days, being Catholic, Apostolic, and Roman, she could not get married again. According to the teachings of the Catholic Church, it was forbidden for a widow to remarry. That ruling was later relaxed to allow widows to marry again under certain conditions. My godmother was devoted to household chores like cooking, washing, ironing, keeping the house extremely clean, and of course, and sewing.

As I said earlier, my dad was a master builder, responsible for many employees. He worked on repairing the walls of El Morro Castle in San Juan. A relative named Toñito showed him how to read blueprints and deal with employees. Then they rebuilt a bridge over the Rio Blanco whose current flowed to Naguabo beach. That bridge still stands today just as they built it.

They worked repairing roads and bridges. At that time, there was no machinery. Everything was done with men's muscles ("pick and shovel"). Carpenters made wooden molds, inserted steel bars, then hand-mixed the concrete. After filling the molds, they waited two or three days, then they removed the molds and what remained were the concrete structures. Many of these structures still stand today.

I remember the days when the sugar cane industry was a vital resource for Puerto Rico. My dad had a farm where he planted sugar cane. A horse turned a mill. In the sugar mill, they squeezed the cane in order to extract the juice that was collected in 55-gallon metal barrels. My dad sold the juice to the sugar refineries where the rest of the process was performed. Besides black, brown and white refined sugar, the sugar cane harvest also produced acetone, sugar cane juice, and rum.

Life for the sugar cane workers was not an easy one. They worked from sunrise to sunset for a dollar. They were exposed to all kinds of weather conditions: hot sun that could crack stones, snakes, **the pica-pica (**a stinging herb which grows in sugar cane fields**),** and wounds from machetes. Early in the morning, they ate a small breakfast of coffee and bread. Around nine o'clock, they had coffee and bread or water or any combination of the three. At noon, there was a small lunch of corn meal, coffee, and maybe a fried fish.

To break the monotony, the workers sang and improvised songs. Many popular songs were born in those fields. I even hear some of those songs played on the radio today. The topics were varied: failed love, regrets, songs to girlfriends or wives, and protest songs.

I remember how pretty it was in the countryside of Puerto Rico. Sadly, much of the flora and fauna of that time is gone, but some things do remain. Coquis still sing in the night; roosters, hens and birds still announce the arrival of a new day, and finally, dawn still breaks over the beautiful Puerto Rican countryside. I will always miss the smell of fresh earth, fruitful and intoxicating, morning mist, fireflies at night, the mooing of cattle.

My dad had a truck and a car at that time. The truck was a 1941 Chevrolet. The car was a 1935 Ford-T two-door. As I was growing up, my father changed the family cars several times. He had many vehicles. The last was a 1948 Chrysler Imperial. Ironically, my dad never learned to drive, but instead always had a driver-employee, who spent the day and perhaps part of the night with him. The driver was also his bodyguard. Because my father did not trust banks, he always carried a lot of money in his pockets. "Bankers are a bunch of crooks," he used to say. His employees were always paid in cash using small sachets or envelopes.

Some Sundays my father picked me up and took me along with my brothers and sisters to enjoy a trip around the Island. Few people were driving at the time so when they came upon another driver, they were respectful, greeting each other warmly. The beautiful panorama of the sugar cane fields impressed me the most. Resembling a white sheet, the flowers of the sugar cane announced an abundant harvest. The workers would have jobs to support their families. They lived in what were called colonies, which were small houses inside the large sugar plantations.

PUERTO RICO GETS OUT OF ITS MISERY

During that era, most of the peasant houses in Puerto Rico were roofed with straw, and the walls were made of palm fronds. The floor was dirt. At the entrance to the house, there was always a "batey" which was highly typical at the time. The batey was the meeting place for the whole family. People sat on rocks or benches positioned in a circle or semi-circle. On the sides were flowering plants such as jasmine, tulips (gardenias), and roses of all colors that embellished the entrance of the house. As there was no radio or television, the only pastime was talking or gossiping until nightfall. Then people drank coffee or ginger tea before going to bed to increase the Island population.

Further away stood a "tormentera" which was a small house used during the passing of a hurricane or an approaching storm. In the kitchen, a three-stone "fogon" was used to produce fire, using dry wood for cooking food. The lady of the house placed pots on the "fogon" to use the heat energy to cook food. They slept in hammocks or cots. The cot was closed during the day and opened only at night. There was a small table with a couple of wooden benches. They had to go to the river to fetch water in a brass container with a three or four gallon capacity. There were no sanitary facilities. The latrine, a hotbed of diseases transmitted by insects, bacteria, and viruses was outside the house.

The absence of suitable sanitary facilities was the main cause of gastrointestinal diseases, particularly affecting children who wandered barefoot around the outside of the house. Even in the slums where I lived, every house had its own latrine. In those days, I suffered from many intestinal diseases from worms and parasites. In streams and other bodies of water, there were all kinds of mosquitoes that transmitted bilharzia and other intestinal diseases.

As the years went by, straw houses and palm fronds gave way to wooden houses with wood floors and zinc roofs. Usually the houses were erected on stumps for a strong foundation. The wood used was resistant to moisture and termites. Then people started acquiring furniture and domestic appliances. Puerto Rico was becoming modernized.

I remember seeing fruit trees such as corazon, quenepas, mangoes, tamarinds, jobos, mamey, and other plants whose fruits or roots were used for food. Other examples are taro, yams, and sweet potato, which were actually part of the roots of certain plants. I should also mention "achiote" that was used to give color to foods. Many people had other medicinal herbs and trees that are no longer seen today. The higüera tree was immensely useful because the dry fruit became hollow containers used to aerate rice and other grains and even liquids like coffee. They also made the famous "dita" from the higüera tree which had multiple uses. With a dita, one could make a "jataca" to draw water from another larger container. Musicians used them to make maracas. These utensils have disappeared from our modern houses. That is the price we pay for modernization.

There was another type of gourd which was more elongated. People used it to make large spoons for serving rice or ground meat. These utensils were exceedingly basic and well known in my childhood. In the back of the house, my mom had a garden which was particularly common at that time in many households. She planted herbs to treat different diseases. People used to cure certain health problems with teas made of those medicinal plants. For example, I remember aloe vera, yerba-buena, rue, eucalyptus, guava, tuatua, chamomile, tau tuba and others.

My mom always had chickens in the back of the house. The hen laid eggs which were darned tasty indeed. The chicken was used to make a number of dishes such as rice with chicken and the famous "asopao." Occasionally she raised a pig. Pork meat was, and still is, useful to make typical Puerto Rican dishes such as sausage, roast pork, rice with pork, chops, ribs, "chicharon volao" and many more. The leftovers were never thrown away or discarded; my mom used them to feed her little pig. In the rest of the Island, people used food leftovers as food for breeding pigs. There were even people who bought food waste wholesale. Even today, many families continue using these methods for breeding pigs. Despite the poverty during that time, people followed this regimen and they maintained excellent health. In my house and other houses, breakfast consisted of any combination of café (black or with milk), bread, oats, cornmeal, and fruit juice. Lunch or dinner consisted of any of the following: flour, rice, beans, cod fish, beef or pork, beef soup, bone beef soup, corn, and plantain balls.

People dressed in clean, ironed clothes. Clothing was never discarded. People covered worn-out clothing with cut out fabrics. One might see people going to church with extremely clean clothing, but here and there you would see the patches on their clothes. It was considered acceptable to dress that way. The most important thing was to have clean clothes to cover your body. Every Sunday, I saw holes in people's socks and all kinds of patches on their clothing, but it did not bother me. I merely considered it the fashion of the day for poor people. Shirt collars and cuffs were starched. Linen was bleached with indigo. The iron used for pressing clothes was made of metal painted black and heated with either wood or coal.

In my neighborhood, there were many people who went barefoot. I was lucky to wear shoes when I went outside. As shoes were a luxury for some people, when their shoes wore out, they would put in a mid-sole to make them last longer. Men even put horseshoes on the heels and front of their shoes to increase their life. Some people did not wear socks; others wore them but kept mending them until there was nothing left to mend. On the Fourth of July and the day of the Three Kings or Wisemen, the men wore a white suit and either white shoes or brown and white or black and white shoes. Small holes in the shoes were painted white using a toothpick. As a male, I focused on what men wore—both their clothing and their shoes. Older women wore dresses with long sleeves. Young women wore high-necked dresses that covered their knees. When a woman sat, she never showed her legs (or anything further up). Showing cleavage and thighs were particularly common practice for prostitutes, and women did not want to be mistaken for prostitutes. Prostitution existed in certain businesses. Each one always had a man at the door controlling the entrance to the bordello. That man was also responsible for keeping out the undesirables. He lifted them like a sack of potatoes and threw them out where they fell like soursop onto the pavement. In my village, there was a business that was called El Danublio Azul (The Blue Danube). People said it was actually a brothel. I do not know if it was true.

As there was no radio or TV in those days, my mom used to tell me stories before bed at night. Popular stories at that time were Johnny the Dummy (Juan Bobo), Red Riding Hood (Caperucita Roja), The Cuckoo (El Cuco), and Snow White (Blanca Nieves). There were people who were experts in the field and charged for reading fairy tales or invented their own. Poems and nursery rhymes were also part of the repertoire.

During the day or early evening, street vendors could be heard coming through the streets selling their stuff, or exchanging oranges for beer bottles. Other street vendors announced their wares such as ice cones, ice cream, tripe, and fish. I remember a man who came weekly, offering to sharpen scissors and knives. There was a superstition that when this guy showed up, he would play his mouth harmonica, and one had to put your hands on your head for luck. The vendor who stretched mattresses was another unusual guy. In those days, there were no spring mattresses. They used a kind of wire frame that needed meticulous tuning after a year or two. There was another man who sold food and vegetables, pots, roasted almonds, shortbread, roasted peanuts, and all kinds of fruit juices. With hawkers coming through the streets of the neighborhood selling things in the early morning or evening, there was no need to go to the market. There were men who came from Arab countries who sought asylum on the Island. They peddled fabrics, dresses, shoes, and other trinkets. Those vendors were remarkably astute, and people believed the stories they told about the goods they were selling—stories which were often totally false. It saddens me that those images of my old barrio exist now only in my mind.

The government's Department of Health had control over sales of meat, milk and fish. All towns had a market square where agricultural products, poultry, and dairy products as well as other things were sold.

In those days, cockfighting was immensely popular, not only in neighborhoods like the one I lived in, but in many towns and villages. The combination of liquor and cockfighting always ended in a tremendous fight where men used either their own fists or armed themselves with knives or machetes. Quarrels among friends or neighbors, cheating on a spouse, or dating your next door neighbor were the principal triggers for a fight. I always heard people say, "that guy has horns in his forehead." I looked at the guy and could not see the horns. As I was extremely sheltered, I did not know what they were talking about. Of course, I later learned that the wife was cheating on her husband. All these fights ended in one of three ways: the hospital, the cemetery, or jail.

Most of the men walked around armed. There was a popular knife called "la sevillana," which had a large curved blade. When they cut someone with that knife, their intestines fell out. If I saw that, I could not sleep. Fighting with machetes also provided an unusually bloody spectacle. There was a man who wore a long iron bar; another had a "blackjack." Old men always walked with canes made of hard wood which proved to be a sufficient deterrent, because they could easily break your bones. Very few people were armed with guns because they cost a lot. The military veterans who served in World Wars I and II always carried either a Colt .38 or a .45 A German pistol. Those bullets were considered deadly. There were people who prepared the bullets with explosives that would break into fragments when the bullet entered the victim's body. These are the visions I saw while growing up. I was determined that someday I would get out of that slum and live a decent life.

In my neighborhood, people played what they called "black ball." The "bolitero" sold numbers and people played to see if their numbers matched. If they won, they might be able to escape from poverty. As there were many Dominicans in my neighborhood, people also played the lottery of Santo Domingo. These were rough expatriates from the regime of the violent dictator, Rafael Leonidas Trujillo, who ruled Santo Domingo with an iron hand. My next door neighbor, Don Genaro, was a Dominican expatriate. He was the only person who owned several beautiful radios, some of which were shortwave radios. He was able to tune into the shortwave radio station signal of the Dominican capital and listen to the Dominican Lottery. The boliteros used that lottery to determine the winning number. The black ball used the last three digits of the winning number.

Many expatriates from Santo Domingo, Haiti, and Cuba fled to Puerto Rico. The flow of people from other Caribbean islands was not something new. It happened also in the early 1940s. Many of these people brought their talents to Puerto Rico. It is true that in any migration, there are trustworthy and rotten apples. Almost every day, I listened to the Cuban program of the famous Guillermo Portables, "Punto Cubano y Sus Guajiras." Rafael Leonidas Trujillo, dictator of Santo Domingo, Prio Socarras of Cuba and Haiti's Papa Doc Duvalier were always in the news, which recounted only the atrocities they did in their respective countries. Years later, another dictator who ruled Nicaragua was in the news. His name was Anastasio Somoza. People called him "Tachito." All these dictatorships were supported by the State Department of the United States, according to the newspapers of the time. To complicate matters, the Americans sent the U.S. Navy to quell any attacks against these dictators. Any uprisings against them were crushed by either the army or navy of the United States. The news media always pointed the finger at the United States. At the time, there was a sector of the Puerto Rican people who were, and still are, against the North American invasion of our land. Don Genaro, my neighbor, used to raise the volume on his radio, blaring programs from the Dominican Republic. People sat in the living room, and the overflow sat in the courtyard of his house to listen because no one else in the neighborhood had a strong radio with such a powerful speaker!

WORLD WAR II AND ITS EFFECTS IN PUERTO RICO

In the early 1940s to 1945, the Second World War was the main topic on the radio and in the newspapers. The Allies were fighting on two fronts—one in Germany that people called Hitler's War and the other one in Japan that people called the War of Hirohito. They had songs condemning the attacks by the Germans and the Japanese. Occasionally at night, there was what Puerto Ricans called "blacao," or in English "blackout." Power plants turned off all the lights, and Puerto Rico remained in total darkness. Then one could see the bright lights of the spotlights from the ground searching the sky for enemy aircraft. The lights came on at the end of the blackout. At that time, we had electricity and running water in my house.

After my Uncle Lolo had died, my mom bought a house for $300. It had a concrete balcony, and the rest of the house was made of wood covered with zinc. The floor was made of wood and was mounted on a base of treated mangrove wood. The latrine was behind the house in the yard. Behind the house, there was also a garden with a breadfruit tree, a guava tree, a cherry tree and an avocado tree. The house had three large rooms. In that house, I spent the best years of my childhood. In the evening, we did not close the doors; the windows were left open. Thus, cold air circulated during the night. There was no fear that thieves would break in.

From the balcony, I could see Zeppelins in the sky. The airships were popular at the time of war. Rafael, my mother's brother, lived on O'Donnell Street in San Juan. Almost every Sunday, we visited him. From the roof of his hotel property, I could see the warships anchored at the docks in the marina area of San Juan. It was extremely common to see American soldiers and sailors strolling along the streets. These soldiers were stationed at Fort Brook. Marines were stationed at "Isla Grande." There was a hospital called "Ballaja," but later on, the name changed to Rodriguez Hospital. In that hospital, they treated the casualties of war. Isla Grande had an airport belonging to the navy, because, at that time there was no air force. I saw many single-engine aircraft landing there. At that time, in San Juan there was a convent for those who aspired to become priests. I remember the trolley parked at the Christopher Columbus Square. I liked to go to San Juan because it felt like being in another world. The sea air and the intoxicating smell of the breeze were extremely pleasant. That was fresh air for me. In my neighborhood, one had to get used to the smell of the latrines. Vendors sold toys for children. What most struck me was the man who sold the multi-colored balloons. There was another man who sold Adams Chiclets for a penny. The flavor of mint left my mouth fresh. Another person sold snow cones for a nickel. There were many flavors, but my favorites were raspberry and coconut.

During the war, there were many sailors and soldiers in the area of San Juan. The brothels were in the marina area. My uncle would not let me **go** that way because it was forbidden for children. In Fortaleza Street, there was a movie house called the "Rialto." There they showed war films. That corner was a popular meeting place for local people. From there, one could see pretty girls walking down the street. That spot faced a Catholic school where my cousins went. I also saw a lot of men making jokes to the girls passing by in front of them. My uncle told me not to visit that area because that was where the lower class people traditionally met. The place was called LA BARANDILLA.

CHILDREN'S HOBBIES

Kids of my age usually played in front of my house, and my mom let me play with them. They had different games such as "bolita y hoyo," "villalba," "blind man's bluff," and cards. We used to make our own toys from whatever materials were available at the time. I enjoyed playing with a "trompo," (a spinning top), a yo-yo or pushing an old wheel bike. We took off the spokes of a bicycle wheel, and we pushed it with a wire hook so that we controlled the wheel. Also, we used the old wheel of a Ford tire, and we pushed it with one bare hand to make it go down the street. We hung swings in the trees and rocked until we got tired. We also played baseball. For the ball, we used an empty thread spool that was made of an extremely hard wood. For the bat, we used a broomstick. We marked the bases by drawing a groove in the form of a square in the bare ground. In those days, the streets were not paved. Sometimes we could not find a bike wheel. In that case, we used the metal band from rum or wine barrels to hold the wood in place. The whole idea was to look around for materials we could use to create a game to play.

We played hide and seek. In that game, one child covered his eyes while the others ran off to "hide." When he finished counting and uncovered his eyes, he had to "seek" at least one other child playing the game. Then the child that was found covered his eyes and the game repeated itself as many times as needed.

As I got older, I invented other toys with cans of condensed milk. The other children were amazed by my creation of cans intertwined with a wire. I was able to get all the cans turning by just pulling one string. When the principal can turned, the others repeated the action. I spent several days studying the works of an old windup clock that I found in the garbage. I am certain that my design skills developed from those experiences and ultimately helped me when I became an engineer. It took a lot of thought to figure out the internal working parts of a grandfather's clock. I did it because I did not have anything else to do.

Sometimes I got bored and then I started solving the newspaper crossword puzzles. One day one of my brothers taught me to play a game they called "ladies." I learned it so well that my brother did not want to play with me because I always won. My older foster brothers bought me toys from my cousin's shop—mostly cars made of wood or rubber. Plastic had not yet been invented. I also bought puzzles to see if I could put them together. These mental exercises helped me to improve my technical skills.

THE THREE KINGS DAY AND OTHER CUSTOMS OF PUERTO RICO

One of the most beautiful traditions at that time was the Three Kings' Day celebration. My mom told me that the Epiphany was a religious festival where it was necessary to go to Mass to celebrate the coming of the Messiah. Jesus Christ was born in a humble manger in a stable in Bethlehem surrounded by a mule and an ox. Three Wise Men came from the East to honor and celebrate the coming of the Messiah. Guided by a star in the sky, they found the crib and brought gold, frankincense and myrrh to the Child Jesus as an offering. My mom told me that the Three Wise Men came on camels. No horses were in Bethlehem at that time. The kids cut grass and left it in empty shoe boxes. The camels ate the grass and the Three Kings left their gifts for us. We put out a pot of water for the camels to drink.

At that time, the parents cut the grass with scissors to simulate the footprints left by the teeth of the camels. As I had a strong Catholic training from childhood, I believed blindly what my mom told me. On January 5 every year, I went to the mountain to get grass for the camels. That day, I went to bed early so I could get up early the next day to look for my gifts.

The next day, my friends and I played to the utmost. It was incredible to experience a time where creativity had no limits. I kept that tradition until I was fifteen years old when I finally learned from my older brother who the Three Kings really were!

I was extremely lucky because every year I received expensive gifts. The first toy cars were made of wood; later they began using rubber. Year after year, things got better. The gifts that had the most influence on my childhood were a tricycle, a scooter, a wagon, a child's bicycle. When I was around 12 years old, they gave me a new 24 inch bike the so call size 24 for boys. Later on as a teenager, I got my first men's bicycle size 26. As expected, the boy's bike was smaller than the men's bike. The best Three Kings' day for me was when I got that size 26 bike. I will always remember my 26-inch "Monarch" bike." It had what they called a "mechanical knee." When the front wheel fell into a hole, it absorbed the shock. Today it is called a "shock absorber." I learned bicycle mechanics by disassembling and then re-assembling that bike. My dad gave me all the tools I needed to fix it. My friends and I went for a ride around the neighborhood. We climbed a mountain then raced downhill at an incredible speed. I was exceptionally skillful at handling the bike. I learned how to get from Rio Piedras to Santurce across Ponce de Leon Avenue. One day we bicycled all the way to the city of San Juan. I went all the way up to the castle which was the residence of the governor. Then I returned from San Juan to Rio Piedras. When I got home, I was exhausted.

Later, I bought three old bikes and started a bike rental agency. I used the money I earned to go to the movies. One day, it occurred to me to start saving to get a better bike. In Rio Piedras at that time, there was a bike shop named Medina. I became a friend of the owner. One day, I asked him to lower the price on a Schwinn Majestic so that I could have a bike that was truly appealing. Later on, I bought many ornaments, lights, bells, clappers, and whistles. Owning that fabulous bike made me the envy of all my friends. But more significant was the fact that, without knowing it, I was becoming an entrepreneur.

Sometimes though we make mistakes. One day, Mr. Medina told me he could sell me a gasoline engine that I could adapt to my bike. I started saving for the engine. Mr. Medina put the motor on my bicycle. At that time, the barrio of Capetillo had recently tarred the streets, and it was usual after they poured the tar, to put on a final layer of a finite stone called "stone dust." One day one of my friends invited me to compete with him in a "little competition" in the Capetillo neighborhood. I accepted the challenge, and we were followed by a bunch of boys who planned to watch us compete. We were supposed to ride 200 meters in a straight line. At the end, we agreed on a motor bike competition of 300 meters on that road covered with stone dust. I do not know what happened, but before I could stop my bike, I catapulted over the handlebars at a speed of about 60 miles per hour. I landed on the pavement, scraping my chest with finite pebble dust. The motor bike suffered significant losses. Fortunately, my mother was not aware of these crazy adventures.

I was in bed for two weeks. To this day, I have not used any "motor bikes" or motorcycles. I had to sell all the bikes, but I kept the Schwinn Majestic frame. With that frame, I put another bike together, and I had it until I graduated from high school.

Returning to the celebration of Epiphany, I think it is a shame that American merchants conspired with local Island traders to change the Island traditions. They replaced them with others that are not of Puerto Rican heritage. Nowadays, we do not observe Three Kings' Day as we once did. The same thing has occurred with other traditions. It might be appropriate to celebrate "Santa Claus Day," but we should give more emphasis to Three Kings' Day than to Santa Claus Day. Only the Islanders can change all that. The business people will probably keep celebrating the Epiphany also in order to sell more products. They do not care that people do not have large pockets filled with money.

Epiphany was not for children only. The adults found a convenient excuse to extend the tradition longer so that they could party more. We are probably the only country in the world where Christmas festivities start right after Thanksgiving Day and extend up to the first week of the New Year. I remember that both in the countryside and villages, people went out partying during January 5, 6, and 7. Then they celebrated the "octavitas" which continued into part of the month of January. It has been said that the consumption of alcohol peaked during that time of the year. That is why, at one, time, we had considerable distilleries on the Island. I remember that during that period of the year, one of my uncles stopped working so that he could party. The jibaros or people from deep in the Island found this time of the year as a "release valve" to get out all their frustrations. As a child growing up, I watched this crazy excitement going on all over the Island. At that time, I could not understand why people were acting this way. That was a tradition year after year until it slowly disappeared. I would say that the people in the sugar cane fields of Puerto Rico had a fantastic time—better than those in any town center. It is unfortunate that customs are disappearing and being replaced by customs of other countries. Newcomers at that time adapted to this way of life. Even people from the other Caribbean islands absorbed these traditions to blend with their own customs and practices.

One day, my dad took me to a river in the little town of Morovis where he was raised. His brothers and others in the family still lived in that neighborhood. We went fishing in a creek where I saw river shrimp, Dajaos, guabinas, eels, and other kinds of fish. Whatever we caught was served for lunch (or dinner as the case may be). My oldest brother, Roberto, and my cousins Tony and Pepe were capable swimmers and could be under water longer than us. I ventured into the water hole until the water reached my shoulders. My uncle Salvador's house was at the top of the hill. We climbed up then descended to the bottom, ending on the river. We sat on a palm leaf or "tigüero," which other people called "yagua." These were products of a palm tree and were particularly useful for sliding down the hill. The danger was hitting any rocks or stones while sliding down. When this happened there was always blood. We arrived at Uncle Salvador's house bruised and bleeding. My aunt threw salt on the wounds, and the problem was solved. Salt was used in the absence of iodine or Mercurochrome. Salt was also used to preserve food. It was a necessity back then. Poor people did not have the money to buy a refrigerator. Some Caribbean islands were the source of salt. My father said that the salt mines in Puerto Rico were unusually scarce.

I liked to go to the countryside of the little town of Morovis. There was some kind of celebration going on Friday or Saturday. The instruments used were "cuatros," guitars, maracas, and guiros. The jibaros played flawlessly without microphones and electronic amplifiers. There, I saw people performing dances such as boleros and "guarachas" that were in vogue at that time. The radio played music. My father was the only guy who owned a portable radio which was the size of a suitcase. The radio had vacuum tubes and a powerful battery. The sound was not perfect because the radio stations were located mainly in San Juan. I remember that they played Cuban music from El trio Matamoros. They also played jibaro music like seis chorreao and mazurkas. Sometimes people sang what they called "bomba." Music from sextets and septets was also extremely popular during those days. The most popular band at that time was the "Septeto Puerto Rico" The sextet and septet music was the collections of Cuban music of the time. The most influential Cuban septets were "Septeto Ignacio Piñeiro" and "Septeto Habanero." In the festivities that were called "Fiestas Patronales," which were in celebration of the patron saint of that particular town, one could see all kinds of music groups, and the different instruments. To get a joy ride we have the Ferris wheel which was approximately ten stories high. There was another ride that people called "El Gusano", which was a ride on a circular railroad track, but that it was going up and down like a roller coaster. From a distance, the thing looks like a serpent or what people called "The Worm" which in Spanish is EL GUSANO. I saw a crowd betting on a game that they called LOS CABALLITOS (the horses). This certainly was wooden horses running in a circle in something that look like a horse race racetrack. My father likes to play those wooden horses games, but, he always lost money. What most impressed me were the dramatic and colorful fireworks. It impresses me the most the rapid rise of what it looks like a rocket being launched vertically upward. When that thing reached the maximum height, it exploded breaking up in colorful tears coming down to the earth and shining the night sky. That is why the drive to the little town of Morovis was invaluable to me. I was able to see all the splendor of the populace enjoying life to the maximum. I should point out that I was able to attend the "Fiestas Patronales" in my boyhood town of Rio Piedras. In Puerto Rico each town adds their own particular flavor to the "Fiestas Patronales."

Holy Week was a week of prayer and sacrifice. At that time, the Catholic Church had much impact on the population. Probably 90% of the population was catholic. We had to do penance and fasting; in other words, no meat and no filling your stomach. People came down from the farm lands to attend mass and other religious activities. Since the paths in the area were extremely muddy, people came down the hills to the road barefoot. Upon reaching the road, they washed their feet and then put on their shoes. In the field, there was no toothpaste. They picked a few leaves of cockle burr and with one finger cleaned their teeth. They grabbed a tin can or basin with water and rinsed their mouths with caution because no one could swallow water as Communion had to be received on an empty stomach. When the Mass was over, they could drink coffee or whatever they pleased.

Palm Sunday, Good Friday and Easter Sunday were the three most celebrated days for the entire population. Country people and city people wore their best clothes. My uncle told me that the jibaros came down to town only on those three days. This was hard for me to believe. Those who were devout Catholics never missed Mass on Sunday. The church announced the Mass with giant bells whose sound could travel for miles. They rang the bells every 10 or 15 minutes three times before the Mass. Women wore veils over their heads; the men had to remove their hats while inside the church. One day, I saw some riders coming to the church on horseback. They tied their horses to planks of wood on the outside of the church. Many jibaros had beautiful horses and saddles adorned with beautiful colors. They were proud of their strong horses and liked to show them off to their friends. The horse was the only means of transportation for those with the funds to buy one. The priests had the best horses because they had to say Mass in chapels far away from the main church.

The priests were Spaniards, and some of them had terrible tempers. I remember them scolding me because I was looking at the ceiling of the church. There were different types of priestly orders—Franciscans, where I studied, Dominicans, Benedictines, Jesuits, and many others.

Walking around the central plaza was customary at that time. The girls wore their best dresses, and some wore pretty shoes. Yes, they were looking for a "victim" as we called it in those days. As progress increased, gradually the number of horses decreased, both on the farms and in the villages, replaced by various types of motor vehicles. I remember seeing trucks, cars, station wagons, buses and so on. The most popular makes were Ford, Chevrolet, Dodge, Fargo, International, Hudson, Plymouth, Mercury, Packard, Nash, Studebaker, Jeep and many more that have passed away into history.

Returning to the subject of religion, I remember participating in the famous Holy Week processions. It was also the practice of people to go to the movies to watch "The Life, Passion and Death of Jesus." During the film, people cried watching the injuries inflicted upon the body of the Lord Jesus Christ. I already knew that movie by heart, so I told my mother that I did not want to watch it anymore.

There were people who replicated the wounds on their hands and side to experience the reality of the chilling crucifixion of Jesus Christ at the hands of the Romans and the Pharisees. People were genuinely experiencing the last days of Christ on earth. I also remember hearing many of the liturgies that they practiced. In those days, people had much more respect for and fear of God. Unfortunately, this has been forgotten, and now we see the consequences of drugs, crime, theft, and all kinds of violence. For some people, Easter is like any other week. I remember that during Easter week, the radio played serious or religious music only every day. On Good Friday people always listened to religious programs. The most famous was the sermon of the Seven Words. I never missed the Sermon of the Seven Words, offered by the Jesuit priests. The Jesuits are religious scholars. To become a member of the Order of the Jesuits, candidates had to study hard. The first Catholic schools in any country, including several universities, are always administered by the Jesuit Order, and they are the finest teachers.

I must admit that, during the course of my life, I have moved away from religion because the flesh is weak and the devil is always lurking there to entice one to hell, but I've always believed in the Catholic Church and go to church and take communion. If people had a fear of God, then there would not be all these hurtful atrocities that we see every day. Of course, this book is not about religion, so I want to stop here.

My mother cooked well, and she was tired of the three stones in her kitchen and using coal. One day she came in with a kerosene stove. In my neighborhood, there were not that many people who cooked with gas or kerosene stoves. I always remember her firing the burner with "Three Stars" matches. Food cooked faster and had delicious flavor. My mom told me I had to be careful not to play with matches because the house could catch fire. Thank God she did not have to take bags of charcoal or go to chop wood in the "cercao." The cercao was a property that was fenced with barbed wire. It was close to the house. I put a sack of potatoes (the material was hard) on top of the fence and jumped to the other side. I chopped the wood and returned home in the blink of the eye. I was quick in my endeavor because, on that farm, there was a strong bull and some scary dogs. Several times I was in danger from the bull, and one day a dog bit my leg. I spent two weeks in bed. But the bull was worse. It pawed the ground, foaming at the mouth, and with eyes flashing took off after me. I had to run for my life. The farm was owned by the Davila's family. We never met the people. I knew the butler (Ulyses O'Neill) because he was also one of my neighbors. I saw him returning to his house every day at the same time. I knew that his brother Don Juan liked me. The woman's name was Juanita. I played with the children of this marriage, Teli, Raul and Junior. Well one day I was picking fruit on the Davila's farm. There were many guavas, "jobos," "guamá," mangoes, and many other fruits. Suddenly I heard a shot in the air, and I almost died of fright. One of Mr. Davila's workers accused me of stealing fruits from them. It was extremely scary because I was just a child. In the distance, I saw a guy running towards me. He was the butler, who was my neighbor, and pleaded with Mr. Davila not to hurt me because I was an innocent child.

Many of the fruits that I saw on that farm no longer exist or at least have been disappearing. Also on that farm there were many birds such as warblers, broad-winged hawks, nightingales, larks, "pitirres," and others. All these birds have also disappeared from the beautiful farm country and villages of Puerto Rico.

That farm had the only river that was left in the city of Rio Piedras. The water was unusually clear. One could look at the bottom of the river and see different colored stones. Unfortunately, the river became another victim of modernization. They rerouted the river in order to build a new road. There was a creek that we called "Quebrada Juan Mendez." Juan Mendez was a celebrity in my neighborhood. This man distributed the newspapers, "El Mundo" and "El Imparcial" and lived in a house facing that beautiful creek. Many children, including me, swam there for hours. Unfortunately, the creek was contaminated with a parasite called bilharzia. This parasite was particularly dangerous, and the symptoms were fever and chills. The parasite significantly affected the human liver. Laxatives were not effective against it. The only thing I remember is that my mom took me to a pharmacy, and there they gave me some shots. I had to get one shot a day to kill the parasite, "Schistosoma," which is its scientific name.

That was not the only parasite I picked up. Going barefoot to the latrine was also another source of parasites. The parasites of the intestines are many, but I remember passing worms from my anus every time it was my turn to take the purgatives. My mom took me to the Public Health Unit of Rio Piedras, and there I got the most popular purgatives: "sal de soda," the castor oil laxative, the purgative "epazote," the purslane purgative, and the "zapayo" purgative. I had to drink all those laxatives because I had worms, tapeworms, and other parasites that took up residence in my intestines. To this day, I can still remember the awful taste of those notorious purgatives and my stomach still get twisted into knots. It is something that I will take to the grave.

MY ELEMENTARY SCHOOL AND THE PUERTO RICO OF MY ADOLESCENCE

I went to a Catholic school, where I was required to make the first Holy Communion. I wore my first and only suit for that occasion. For my foster brothers and me, that was an exciting event. They bought me a rather delicate white suit, white shoes and socks. They also hired a professional photographer to take pictures that I still keep, despite many years that have passed.

In San Antonio School, all students had to wear a school uniform. The official uniform for children was a short-sleeved white shirt, brown shorts, brown shoes and socks. I had to be well-groomed and had to take a bath every day. Girls also wore a white blouse, brown jumper, socks and brown shoes. The first Friday of the month, we replaced brown with white. For children receiving their First Communion, going to confession was required, and they had to receive communion every Friday.

The nuns spoke English well and maintained a strict discipline. I learned to speak English from the first grade, increasing my English skills every year thereafter. I learned to speak, write and understand the language correctly. That was a blessing for me. The nuns punished us if we violated any of their rules. This disciplinary atmosphere bothered me a lot. One never knew when punishment might be meted out. It was impossible to remember so many rules of conduct and behavior.

When I was a kid, I always behaved well with those around me. I saw no reason to be angry or throw a tantrum in public. My cousins were owners of LA RIVERA store and LA COLOMBINA store in Rio Piedras. My father had built those buildings. When I went to my cousins' stores, they would give me clothes and toys. My dad gave me money when he saw me. I never lacked basic clothing.

I knew that my dad drank rum. When he saw me, he gave me a nickel or a dime, which for me at that time, was a lot of money. When I was only 8 years old, he took Coca Cola cap and used it as a container. He poured rum into the cap and gave it to me rum. Then he said; "Drink it." The cap could hold only probably an ounce of rum. I drank the rum then spit it out because it was unusually strong for me. I just took a few drops of rum. Then I remembered that he told me, "Never drink rum or beer. You must wait until you reach your 18 birthday." The legal age to drink was 18. However, I did not heed his words because after that, during Christmas day, my friends and I drank the alcohol that other people left at the tables. Of course, we kept it a secret. I must admit I only drank less than an ounce or two of rum. I was terribly afraid of wine. The consequence of drinking cheap wine sold at the time was the worst thing that could happen to anyone. During Christmas, we saw all those people drinking, dancing and eating. My mom loved to dance. Many musicians came to the house to party. They always played jibaro music. Also, they played Mexican music like "rancheras", "corridos", and other popular songs, especially those of Jorge Negrete who had visited the Island of Puerto Rico. People sang tangos because Carlos Gardel, a famous Argentinian singer, came to Puerto Rico in the early 30's. After he left the island, he was killed in an airplane accident in Medellin, Colombia. The tenor Jose Mojica was also extremely popular on the Island. Another Argentinian singer, Libertad Lamarquez, was tremendously popular, especially among women. A lot of Argentinian songs were sung. It was also customary to sing "Alla en el Rancho Grande" because its composer, Tito Guizar, was in Puerto Rico for the premiere of the movie of the same title. Mexican films of that time had much impact on the folklore of Puerto Rico. The cry of the Mexican charro was imitated by all those people who sang in my house.

In my neighborhood, there was a toy store owned by a man named Don Angel. I bought revolver caps then played cowboys with my friends. Comic magazines became extremely popular at that time. My favorites were the adventures of Roy Rogers and Gene Autry. Up to my twelfth year we played cowboys games. At that time, my mom let me go to the movies alone. In Rio Piedras, there were several theaters ("Atenas," "Modelo,"" Lux," "Paradise" and "Victoria") that charged a dime during the week and a quarter on weekends. There was another that was called "el meaito" because the people pissed on the floor since they did not want to miss the excitement of the movie that was being shown. That theater was called "Hatuey" and was just off Rio Piedras on the way to the town of Carolina. Because the owner was a bottle collector, one could bring in one 16-ounce empty bottle of Indian beer and one get in free; otherwise, one have to pay ten cents. Then they started raising the prices. Instead of a dime, they charged 15 cents and then went up to a quarter. People that pissed on the floor were thrown out like a sack of potatoes. People stopped going to that movie house, and they had to close the place. The worst problem with this place was the unbearable stench of urine. It also became known as the meeting place of the underworld, which met there to discuss business. All movie houses were equipped with air conditioner, so people were not allowed to smoke. The only place they allow to smoke was in the back of the theater.

Besides Roy Rogers and Gene Autry, other cowboys such as Bob Steele, Bill Elliot, and Hopalong Cassidy were especially popular. In Rio Piedras, there was a swarthy boy named Abraham who was half mad. He went to the movies all day, watching each one two or three times. Then he stood in the street near the Hawthorne school of Rio Piedras, speaking in a cowboy jargon, his fists raised in the air as he recreated the fight scenes from the movie. People gathered around him and started applauding. We watched him and laughed at his show. I think that he was not mad because he liked people looking at him. I believe, that for him, that was a kind of entertainment. Tarzan films with Johnny Weissmuller were hugely popular. I saw them all. The Lone Ranger movies and other cowboy movies were brilliant.

By the end of the Second World War, Puerto Rico radio became popular and people started buying table radios. The stations that I liked at that time were WKAQ, WIAC and WAPA that were situated in the San Juan area. Another radio station, WENA, was in Bayamon. Later on in Rio Piedras they had a radio station named the WWWW. The WKAQ radio station programs that were my favorites were Tarzan, the Lone Ranger, Tamacun, and Counter-intelligence. These programs were recorded in Cuba or Mexico and then passed through Puerto Rico. They could be heard from 4:00 p.m. to 6:00 p.m. From 6:00 p.m. to 7:00 p.m. was the news. From 11:00 a.m. to 12 noon, the favorite program was the "Don Rafael Quiñones Vidal Show" with his son Quinonez Borras. It was originally called "The Pre-lunch Hour." Eventually, they changed the format and "Tribuna Del Arte" was formed and moved to the evening. That show was a success because it opened the doors to people who aspired to be singers or artists. The phrase "the flying quarter" came into vogue. That was the winning prize. The quarter was symbolic of meritorious candidates. From that program, many future Puerto Rican artists got their initial push to stardom. Later on, other similar programs appeared, but the most famous was one was called "Radio Revista Sabatina." I liked that program because it had fewer well-known professional artists. The presenter was called "El Colorao Almodóvar."

In Rio Piedras, there was a mill to grind corn for food to feed chickens and other types of birds. The mill consisted of a rather large wooden wheel driven by a gasoline engine. I enjoyed watching the process of grinding corn for the birds. The flour was collected in a large tight container. In the marketplace, there was a section where they sold chickens. People came and chose a chicken. They give it to the owner who twisted its neck, and the poor hen fell on the floor jumping until it stopped moving. There was a cauldron of boiling water into which they put the hen. That softened the feathers, and in a moment the process was done. Then it was wrapped in wax paper and handed to the customer. That cost from one to two dollars depending on the weight of the chicken. These hens ate the corn from that mill. The food was fresh, and mom made rice with chicken, or **"arroz con pollo"** and we ate the leftovers the next day. The vegetables and fresh meats were also brought from nearby farms. At dawn, one could see the multitude of traders and transport vehicles on their daily journey. My grandfather had a booth there, and I usually asked him for a better price. My grandfather's name was Luis; he became a merchant although he never went to school. I don't know how he did the math. Usually, the amount was wrong, but if he made a mistake, the error was always in his favor, and if the customer protested, my grandfather pulled out a sharp machete and waved it in the air. The client remained silent and continued on. My grandfather was a man who liked to chew snuff yarn. Gobs of snuff and spit would sail into a corner. I remember that the spit was extremely dark. He also liked to smoke Cuban cigars, the relatively long ones. He smoked them to the end. He did not like smoking cigarettes. He enjoyed chasing women, usually the single women, and always gave them beautiful compliments, and they went away extremely happy. My grandfather told me that he always respected married women. One day he told me, "Never shit where you eat." He also liked to drink rum. The rum was called "Puerto Rican Moonshine."

My grandmother made sure that my grandfather was very well fed. She was a woman who had 14 children in her only marriage. She cooked well. She had a job making "pasteles" and cassava dough. Also, she was an expert making Puerto Rican sausages and fried cod. She sold all that stuff in my grandfather's business. To support 14 children, one had to work hard. As I said before, my actual birth mother was "Benigna." She was the eldest daughter of my grandparents. The marketplace was the center of trade in neighboring towns--Carolina, Canovanas and Rio Grande on one side and Hato Rey on the other side. In addition, we have to include Trujillo Alto, and Cupey. So business was booming after the Second World War was over. Also included were all the neighborhoods of this large city where I was raised. Rio Piedras was a township and a historic town on the map of the Island at that time.

As I said earlier, in 1944, I was attending San Antonio School in Rio Piedras. All of my family was devout Catholics. I was accepted at that school without paying a penny. In return, I and my family had to help with the necessities of the parish. My foster brothers, Carmelito and Manuel, were altar boys. They attended the San Jose school which, prior to that, was a military academy for boys. They only went there for a couple of years.

"Colegio San Antonio" (CSA) was founded in Rio Piedras in 1928. But, I attended this school from 1944 through 1956. Initially the director of the College was Father Gilberto. Then later on came Father "Juanito" and the last one I remember was Father Claire, which we called him "Padre Claro." The Director of the College was overseen the job of the Franciscan Sisters of Pittsburgh whom came to the institution to run the school. The fathers belong also to the Franciscan order, but, we called them PADRES CAPUCHINOS because they wore a hood on their head sometimes. The hood was part of their garment.

I was responsible to stay after class every day to help the nuns. I made sure that the room was cleaned and mopped. Other times I went to the main office to help as a messenger. That was the price I paid for my education. There was a nun who disliked me for some reason. Let's say that her name was Sister Maria. She tried to humiliate me in front of other students, and I would not let her do that to me. I told her that I was not her slave and that nobody was going to treat me like a slave, and even though I was a poor boy, I deserved to be treated the same way that she treated other students. I understood that my honest work was the price I had to pay, but I expected her to respect me. After "swallowing the bitter pill," she never answered me; she only looked at me with blazing eyes.

There were other nuns who were extremely devout and darned talented educators. The priest in charge of the congregation was Father John (Juanito). There was an assistant to Father John and the rest of the priests. He was also in charge of ringing the church bells. He was so proficient at playing the bells that they could be heard for miles. There was a gardener who kept the gardens and the landscape manicured with a lot of dedication. His daughter was also one of my classmates.

After returning from school, I got home, put the books in my room and I went out to play with the other children. I never did any homework. However, I had an excellent memory and was skilled at mathematics and sciences. I always got "A's" in mathematics (I was the best mathematics student) but "C's" in conduct. I was unusually anxious and a bit mouthy. For nuns that was not acceptable behavior. I honestly did not bother to get "C's" or "D's" in conduct because that was an arbitrary grade on the part of the nuns which neither subtracted nor added anything to my knowledge. I was not very competent in Spanish class. But I did acceptable work in English, sports, art and music, math and science. Again I was insulted by Sister Maria. Let me explain. She was going down the stairs, and I was behind her. Suddenly I observed that her veil was gliding down the steps. Then I jumped and stomped on her veil, which dropped to the ground, exposing her hair. She quickly turned around, slapped me on the face, and sent me to the principal's office. The punishment was to write 1,000 times: "I will behave from now on like a good boy." They threatened to expel me. The principal sent me home and told me that I could not return to school until my parents showed up. After meeting with my parents, they would determine what to do with me. Later on, they told me that the final decision would come from the Director of the institution (Father John).

I explained to my Dad what had happened. Since he was extremely wealthy, and he used to make several donations to the parish, he had better leverage to solve my expulsion problem. The next day, my dad spoke to Father John seeking a solution.

On the other hand, months before, the brotherhood of the Servants of Mary had asked the pastor for a statue of the Virgin Mary to put in a corner of the altar. The statue cost a lot of money because they had to order it from Spain. In short, my dad told Father John that he would give the money to buy the statue of the Virgin. Apparently that helped to solve my problem in school. The principal called me to her office. She explained to me that I could come back to school, but I had to write the following 500 times: "I promise to be a good boy and respect the nuns from now on." Well I started writing the 500 sentences, and I promised to behave. That lasted through the end of the school year. At the start of the next school year, I forgot about the promise. My grade in conduct came back to "C" until I reached the eighth grade. Yes, I misbehaved, but my classmates did not dare to expose me because they knew that deep in my heart I was a talented kid. I owed a lot to my classmates because they never betrayed me. Sometimes they were also punished for keeping quiet about my misbehavior or any other kid's misbehavior. There was Masonic Camaraderie among us.

Finally, came that beautiful day in May 1952, when, together with my other classmates, I had the honor and joy to graduate from eighth grade. My parents bought me the best white suit that was in the store. The white linen suit was called a "ONE BUTTON SUIT" because it fastened in front with a single button. They hired a professional photographer to take pictures that I still have. I also keep my eighth grade diploma in my office along with an album of the graduating class. Looking back, I would say that I was an extraordinarily lucky person to reach that achievement. Those eight years at San Antonio High School gave me the necessary tools to ensure my future success. I realized that Sister Maria was trying to help me. She just used the wrong approach to the problem.

As a gift, my dad took me to Morovis, to his childhood neighborhood, where Uncle Salvador lived. It was a Palm

Sunday. That day, my blood brothers and mother Benigna accompanied us for that joyous event. My father's employees joined the group too. Of course, Daddy's driver, who also played the role of his bodyguard and buffoon, was driving my father's truck.

Down the non-paved roads, I saw oxen carts with their heavy sugar cane loads. I should also mention that oxen were used at that time, to plow the farm. A set of two oxen controlled by the Puerto Rican Jibaro was making furrows in the ground. Another Jibaro was throwing a few seeds. At that time, one could plow one piece of land in a single day. If the farm was bigger, then it would take longer. The thing that that struck me most was that even after a long day's work, the Jibaros gathered in a store, owned by one of my father's relatives, to drink and have a good time. The Jibaros were bare-footed and wore white trousers, white shirts, and a dagger or machete in their belts. They spent their money drinking cheap moonshine or white rum named Ron Llave. They also chewed "snuff spinning;" others smoked snuff made there in the farm. When they went to their houses, I could see them zigzagging and tumbling because of their drunkenness. The next day, they were back working on the farms. In that region, coffee and tobacco were harvested. One saw several tobacco ranches and a few open floors where they put the coffee to dry under the sun. After the coffee was dried, they use a pylon or mortar to break the seed shell. The process to remove the speck of coffee was the next step. The seeds had to remain without the shell in order to dry. The coffee was roasted and then ground in a mill. The smell of coffee was fragrant and traveled a considerable distance. That was the best coffee I've ever had in my life, i.e. made in the farms of Puerto Rico.

When there was a rainstorm, things got difficult. The land became muddy red clay, and people and oxen slipped as if there were soap on the ground. On my uncle's farm, there was a river that was difficult to cross when the water rose. Sometimes there was no alternative, if, for example, an injured or sick person needed transport to the hospital. There was a man called "Tarzan," who was the only one who dared to challenge the river. In order to help people cross it, they climbed on his back, and he swam to the other shore. Others preferred crossing on his shoulders if the water would allow. He was paid in return with what people could afford--some gave money, some gave a bunch of bananas. I cannot forget this guy because to me he was a hero. Everyone loved him; everyone admired him for his courage. Sad to say, one day something terrible happened. A storm was coming, and the river was growing by leaps and bounds. The strength of the current was strong. Poor Tarzan and his passenger drowned. They found them three days later about two miles downstream, their bodies shattered by the blows of the stones. That night, I could not sleep. Tarzan's funeral was a tragic event in that region. Thereafter, no one dared to challenge the river. With the technology we enjoy today, I guarantee it would be easy to build a bridge over the river, but, as I have never gone back there, I don't know if they ever did.

MUSIC IN THE 40s AND 50s

As I said earlier, one of my neighbors, Don Genaro, had a shortwave radio. At night, he searched for stations in Cuba and Santo Domingo. Let me remind the reader, the time frame was in the early 40's when people were concerned about Hitler's invasion of Poland and the Japanese attack on Pearl Harbor. I sat in his living room listening to the world news. To be honest, I could not comprehend the magnitude of the problem because I was a child. Just looking at the faces of the people, I assumed that something bad was happening. At that time, the radios worked with vacuum tubes. His first radio had about 20 tubes, and it was one of those "Crosley" antique radios. The radio had a large horn about 15 inches in diameter. Don Genaro had a long antenna on the roof of his house. The antenna was a long copper wire that ran the entire length of the house. Don Genaro told me that in order to tune in a distant radio station, he needed to have a decent antenna. The tuner was another vital item. I remember that it was like an eye that closed and opened. When the station was properly tuned, the eye was closed. He listened to stations from Cuba and Santo Domingo. The Cuban orchestras were extremely popular at that time. The sextet and septet groups were going out of fashion. The only thing I liked was "El Conjunto de Arsenio Rodriguez." He was known as "The Wonderful Blind Man." The only trio that I can remember was that of Miguel Matamoros, better known as "Trio Matamoros." I enjoyed the music coming out of Cuba especially "La Sonora Matancera," "Orquesta Casino de la Playa," and the "Trio Matamoros." In Santo Domingo, the only orchestra that I enjoyed was "La Orquesta de Papa Molina." That orchestra had the best musicians. They played mostly merengues.

My mom noticed that I liked listening to radio music. She bought an RCA Victor table radio. It was one of the first radios that came with only five tubes. It was an extraordinarily powerful radio. The main disadvantage was that I could only tune in local Puerto Rican stations in the AM (amplitude modulation) band. I would have preferred a shortwave radio like the one owned by Don Genaro. A short wave radio was worth the money.

The most popular band in Puerto Rico was the "Rafael Muñoz Orchestra." The band had many singers, but the best one was José Luis Monero. I also enjoyed Mexican and Cuban music. The CMQ radio station of Cuba was unusually strong. They exported their programs to Puerto Rico and other countries. At the time, they did not have a local competitor. The local recording labels were still in diapers since Don Emilio Azcarraga of Mexico had complete control of the radio stations in Mexico and the Caribbean. Many record companies under the control of Don Emilio also controlled the trade in Puerto Rico. We were a captive market of Cuba and Mexico. American music was not particularly popular at that time. However, there were followers of a small Jazz band.

Many of these local orchestras played at the Escambron Beach Club which was the preferred dancing spot at that time. Many music groups and orchestras traveled around the Island to entertain people. They went to private clubs located in towns in the interior of the Island. These private clubs were accessible only to rich people. At that time, the Spanish did not mingle with the Puerto Rican lower classes. But some rich non-Spaniards were able to penetrate into the cream of Spanish society. Meanwhile, poor people had formed their own social clubs sponsored by the Catholic Church. Some people did not have the funds to pay for a renowned orchestra. That situation opened the opportunity for new music groups to play in the "Fiestas Patronales." These fiestas were celebrated in each town to commemorate the anniversary of the town's Patron Saint. Many singers, ensembles, trios, and artists came to those festivities which then prevailed inside and outside Puerto Rico. For example, in the town of Mayaguez, the William Manzano Orchestra was extremely popular. The band was originally called Orquesta

Hatuey de Mayagüez. Years later, it was known simply as the Orchestra of William Manzano. I do not know why the band never succeeded throughout the Island, but years later, many singers from that band achieved international success.

In Puerto Rico, the Mexican custom of "malinchismo" is widely practiced. That is, local residents pay homage to outsiders while ignoring the values of Puerto Rican composers and artists. That is more or less what happened to the "Trio de Johnny Rodriguez". I remember seeing this trio in the "Fiestas Patronales" of Rio Piedras singing their popular hits, such as "Lagrimas Negras." Johnny Rodriguez was the composer of the Popular Democratic Party theme, "Jalda Arriba va cantando el popular." Johnny wrote that hymn at the request of Don Luis Muñoz Marin, who was the leader of the Island for many years. I will never forget "El Trio Borinquén" of Don Rafael Hernandez, whose compositions made history around the world. I remember "El Cuarteto de Pedro Flores" and "El Cuarteto Marcano" which were extremely popular. The singer Piquito Marcano had several songs on the hit parade. Another famous quartet was that of trumpeter Placido Acevedo and his" Cuarteto Mayari." His group had many songs on the hit parade.

At that time on Sundays, it was customary to watch the military and municipal orchestras playing outdoors in the town square. It was mostly poor people who gathered there to listen to free music. There were several Puerto Rican orchestras, such as the famous "Rafael Alers Orchestra" that played "danzas." This genre was considered semi-classical music. The music that I liked was cheerful and lively music. The boleros did not hold my attention. I was too young.

Near my house, there was a "cafetin" that became popular. The owner was the famous Doña Tomasa. She was a widow who had several grown children who helped her in the business. She had tremendous business acumen. There, I ate "cuchifritos", fritters, fried fish, and other fried foods. She also sold well-chilled mabí, which is a Puerto Rican juice. I also loved the rich "mortadella" sandwiches that she made.

One day she expanded her business by installing a jukebox. That was the first and most beautiful jukebox in the neighborhood. It played the old 78 rpm seven inch records. On Fridays, Saturdays and Sundays, men gathered to drink and listen to music. That jukebox was like a piggy bank in which people deposited their salaries one coin after the other. It was the only jukebox player in the barrio. I no longer had to listen to the radio. I just sat in a rocking chair on the balcony of my house to listen to the music of "Panchito Riset", "Trio Matamoros" and "Nico Saquito" repertoire. Nico had been perhaps one of the most prolific Cuban composers. For example, "Compay Gato," "Me tenian amarrao con Fe," "Maria Cristina," "Al vaiven de mi carreta,"and "Meneame la cuna Ramon" and more were played daily on that jukebox. The pleasure of hearing all that music absolutely free from the balcony of my house was quite a lovely luxury. I felt so happy.

One day I heard music coming out of Doña Tomasa's jukebox. I was blown away. They were playing the first mambos of Perez Prado and his Orchestra. My favorites were "Que Rico Mambo," "Caballo Negro," "Mambo Number Five," "Mambo Number 8" and others. Then about two or three months later came another hit. This time the famous Cuban singer Benny More was the main singer in the Perez Prado Orchestra. The Benny More mambos that I liked the most were "Ana Maria," "Anabacoa," "Maria Cristina," and many more. I became so hooked on the Perez Prado Orchestra with Benny More that I started buying their records. My mom bought me an RCA Victor "Vitrola." It was a pretty turntable. From that point on, I started collecting 78 rpm records. My mom gave me a dollar for cleaning the patio on weekends. My dad, who was rich, gave me five dollars. I spent all that money on records, mostly those of Benny More and Perez Prado. I should point out that these records were 78 revolutions per minute with one song on each side, i.e., two songs per record. The records were made of plastic, so they did not last long because if they fell on the floor they ended up in pieces. Furthermore, streaks and scratches affected the sound quality of the recordings. I took care of my albums, but there were always accidents. Also, my cousins stole my records. I lent my records, and they never returned them.

Talking about the music of Puerto Rico and the Caribbean without mentioning the African roots that influenced our music from the twentieth century until now would be almost sacrilegious. To understand the roots of our Antillean music, we have to refer to the Aboriginal group that brought the world the origins of that music. For that reason, we must go back to the past and learn how this phenomenon came from the Islands of Cuba and Haiti, where there are still descendants of these indigenous people.

To explain the presence of the African roots in our music is relatively easy. Through the 400-odd years of the slave trade, black immigrants preserved their religion and African traditions. The Spaniards enacted a law authorizing the Negroes to create "their own councils." These were, and still are, mutual social and welfare assistance associations uniting Africans into different groups, according to the region of Africa where they came from. In this way, there were, and still are, several groups in Cuba. Among the most notable are the Lucumis (Yoruba), the Arara, the Carabalís and the Congos Reales (Angolan).

From these African cults evolved the music which we call today Afro-Cuban music. The fact that these groups were converted to Catholicism did not prevent them from preserving individual rights related to their original religious beliefs. Each council was protected by a patron saint of their choice. Each council believed in the sum of their African gods, plus the Catholic god, and their patron saints. This mixture is known as "**sincretismo,**" an incoherent amalgam of Catholic beliefs of one monotheistic god, and the pantheistic African idea of many gods. In the eyes of Africans, the Catholic god was just one more within their system of many gods. In reality, there was and has never been a real fusion of both worlds.

There is a difference between the Catholic god and the African gods. The Catholic god does not like music or dance. The African gods, who love music and dance, enter the bodies of believers who, possessed by the gods, enter into a trance. During the trance, they shake and twist on the ground. This should not be confused with exorcism. Only the African priest, the "babalao" can make the proper interpretation of what is happening to that person. The "babalao" will decide what to do next. The Catholic priests did not know that blacks were worshiping their African gods or the syncretism in the Catholic religion. For example, when blacks worshipped St. Peter and St. John the Baptist, they were actually worshiping "Oggun" and "Sarabanda."

As another example, "Obatala" was the goddess of the sky (Ajosi of Arara), creator of the world; she was white, and her syncretism was the Virgin of Charity. Ajosi (Obatala to the Yoruba) was an unusually dark skinned woman (Black Madonna), dressed in an immaculate white suit. In her hand, she bore a whisk. They were the creators of the universe. "Chango" was the god of virility, thunder, lightning, and the Catholic syncretism was Santa Barbara. When I was a kid, I remember that blacks from my neighborhood met in a spacious yard and interpreted many rhythms with percussion instruments. Since they did not have money to get professional instruments, they used old furniture drawers, spoons, forks, pots and pans. It was a memorable event. In Puerto Rico, there were blacks who had come from Cuba, Haiti and Santo Domingo and brought their beliefs, customs, religion and music with them.

In my neighborhood, it was common to see groups playing and dancing in the streets to the rhythm of "plena." This rhythm came from the Barrio San Anton in Ponce. The "plena" and the "bomba" were typically heard at their parties. My neighborhood became full of intense rhythms formed by black groups during certain weekends. I could not fall asleep because of the sounds, so I jumped through my window to watch the festivities through a hole that I made in the wall. My mom did not know that I was outside the house.

In 1886, the abolition of slavery occurred in Cuba and other Caribbean islands, including Puerto Rico. The freeing of more than a quarter of a million black Africans created a problem for the islands of Cuba, Haiti, Santo Domingo and other West Indian islands. These people did not have money or own property. Most blacks moved to the periphery of the large cities looking for work, food and shelter. This population movement caused the emergence of many black neighborhoods. It was in these neighborhoods that the "rumba" was born. This music came from the African religious ceremonies.

There are various forms of rumba. For example; the large drum is called "the female," the medium drum is called the "male," and the small drum is called the "fifth." There are several types of rumbas: "jiribilla," "Palatina," "reseda," "mamabuela," "marunga," and "yambú." The forms that have been most popular are the "yambú," "guaguancó," and "Columbia." The Columbia is faster than the "guaguancó;" the" Yambú" is softer. I saw many blacks in my neighborhood playing this kind of music, particularly on weekends.

There is an instrument that looks like a Spanish guitar but is actually called "el tres." The tres have three double strings for a total of six strings. This instrument took center stage of what was known as the "Conjunto Tipico Cubano." When I was growing up, I remember watching many of these tres players interpret the fiery Cuban music either in sextet or septet format. My favorites at that time were; "el Negro Zapatero," "el negro Berto," "Babu," "Mario Hernandez," and one guy they called "Tito El Loco." I would say the best "tresistas" of Puerto Rico came out of the barrio of Rio Piedras where I was born. All weekend long, these people played in sextet format, recreating the music coming out of Santiago de Cuba. The "Sexteto Puerto Rico" was one of the most famous of the time.

Cuban music has been influential in international music. This shows that music has no boundaries, has no flags and is apolitical. One of the most popular songs is "Babalu" which is an evocation of the god who cures diseases; its syncretism is "San Lazaro." This song was written by Margarita Lecuona and was introduced in the U.S. by Miguelito Valdes in early 1950. I've been lucky because I've seen the development of that type of music from my childhood to the old man that I am today. Music for me has always been a vital part of my being. During my life, I have followed the careers of many artists in the field of music, including performers and singers. Most people do not know that "El Bolero" is of Cuban origin. The confusion is based on the idea that Cuban music is synonymous with blazing fast music. To speak of the bolero, one could write a book on that topic alone. Find out the name of the first bolero and who the writer was; then one will discover the true meaning of the BOLERO. I am not going to tell anybody. This is your homework for today.

Since the beginning of the 50's, I began to hear music on the radio from trios' ensembles. I was not interested at first in this type of music. I knew that the music was extremely romantic, and, as a kid, I was looking for fiery rhythms. To complete the list, let me mention some of the trios that came out at that time: Trio Borinquén of Rafael Hernández, Trio Matamoros of Miguel Matamoros, Trio Johnny Rodriguez, Trio Los Panchos, Trio Vegabajeno, Trio Los Romanceros, Trio Los Borincanos, and Trio Los Antares. I would say that the 50's was the beginning of the most beautiful trios in Puerto Rico. Of course, I heard all that music, but I did not fully understand the message because I was too young at the time. Explaining about these trios in detail would be impossible for me. I would have to devote an entire book to honor each of these and other notable trios and many others that followed years later. Of course, I never thought I'd see the day that I would meet some members of these trios personally. I will talk about that later in this book.

NATIONALIST REVOLT OF 1950 AND PEDRO ALBIZU CAMPOS

The Nationalist Party, under the leadership of Don Pedro Albizu Campos, was a group of defenders of the ideal of independence, by force if necessary. I was lucky that in the years 1945 to 1950, I witnessed these men marching like soldiers on a farm near my house. We called them the men in Black Shirts because they all dressed uniformly in black pants and black shirts. I read in the newspaper that Don Pedro had been a colonel in the U.S. Army. The paper said that he was the right person to prepare its armed forces to seek independence from the USA. One of my neighbors, "Don Mode" (his first name was Modesto), was a true nationalist. He used to paint furniture as a trade. He was a Negro man and spent his time lecturing in favor of independence. As a child, I did not quite understand their reasoning. I asked myself why they were always armed to the teeth.

I remember at a rally that Don Pedro said it was necessary to establish the Puerto Rican national consciousness. The people had the right to fight against the Yankee invaders, who occupied the land of Puerto Rico. The Yankees believed that they owned the Island. Don Pedro pointed out that Puerto Rico must be a free sovereign nation. He preached that the people must have Puerto Rican citizenship and abandon U.S. citizenship.

Years later, I had a better understanding of Don Pedro's political philosophy. I dare say that Don Pedro was a patriot who was ahead of his time. At that time, people were more concerned with the financial mess that had caused the Second World War. Using the word "war" when people hated war was not a widely accepted idea. The majority of the Puerto Rican population was followers of Don Luis Muñoz Marin. On the other hand, many people believed in Don Pedro's message. The nationalists and "independentistas" saw Munoz Marin as a traitor to the motherland. The nationalists believed that Munoz Marin had to be overthrown and killed. I think that was the basis for what became known as "The Nationalist Revolt of 1950." I have read the newspapers and listened to the radio every day since I was a child and absorbed all that was happening around me.

Nationalists attacked La Fortaleza, the residence of the governor of Puerto Rico. The newspapers reported that there were many dead and wounded. The patriots tried to reach La Fortaleza Castle, shooting their arms as they approached the gates. The nationalists tried to secure other municipalities in Puerto Rico by force, but the only city that fell to them was the town of "Jayuya." Most of the nationalists were arrested and sent to prisons in the United States. Don Pedro Albizu Campos was taken to Presbyterian Hospital in Santurce, Puerto Rico. The newspapers said that he had been injured and that Muñoz Marin had ordered him to be sent to Presbyterian Hospital for observation. There were two FBI agents stationed at the door to his room 24 hours a day, seven days a week. Pedro Albizu Campos spent many years on his hospital bed. Nationalists proclaimed that Don Pedro was a prisoner of war. There were many protests trying to free him. The nationalists continually tried to get world opinion against the status quo by asking to free their leader. It is terribly sad that he died years later in that hospital. The rumor was that he died of cancer. The nationalist party believers lost a prominent leader. We are still debating the same issues that were debated during those days.

As part of the nationalist struggle against the United States, some nationalists attacked Blair House in Washington, where President Truman was staying at that time. Several nationalists opened fire in the Congress of the United States, wounding five representatives. The nationalists that survived the attack were sent to federal prisons. It was reported that the group had been infiltrated by federal agents. The attempt failed because the agents had already been on the scene before the firing started. In those days, it was customary for agents to infiltrate the enemy. That practice continues today.

It was published that the police were waiting for the nationalists when they attacked La Fortaleza. In these activities, there is intelligence and counter-intelligence. One doesn't know whom to trust.

Months after the nationalist revolt, Muñoz Marin enacted a law to disarm the people of Puerto Rico. I remember that in my neighborhood, police went house to house doing an inventory of weapons and weapons' owners. In my house, there were no weapons, but the police entered and rummaged around the house. In Don Mode's house, they say that they discovered machine guns, pistols and revolvers. I remember that he was taken out of his home in handcuffs. Somebody told me that he died in prison. The 1950 Arm Law only allowed guns to merchants. Those people who could prove that they were not revolutionaries were allowed to bear arms. No one could walk the streets of Puerto Rico displaying a weapon. If a person had no license to carry arms, he was arrested. That was in the days of Muñoz Marin. Today is quite different. It seems that everybody is armed. It is truly sad. You never know when a stray bullet could hit you. Many innocent victims had died needlessly.

In an effort to refresh the memory of many readers, I need to discuss very briefly the history of Puerto Rico. 1898 was the last year of Spanish rule in Puerto Rico. The Spanish-American War ended the 405 years of Spanish rule on the Island. As a result of the agreements signed by Spain in the City of Paris, France, the United States took as spoils of war the following countries: the Philippines, Guam, Cuba, Santo Domingo, Puerto Rico and other small islands in the pacific sea. The United States decided to allow complete freedom to Cuba, but it retained Guam, the Philippines and Puerto Rico as North American possessions. They gave autonomy to Santo Domingo but retained troops on Dominican soil just in case they were needed. I should mention that Haiti is not a possession of Spain or any other country. The Philippines obtained sovereignty but also retained also American troops on the island. The same thing happened in Puerto Rico. The governor of Puerto Rico was a military commander, appointed by the President of the United States. He was not Puerto Rican. There was a legislature under the tutelage of the Americans. This simulated a local autonomy that no one understood. The colonial status of Puerto Rico has not been resolved as of this writing. Puerto Rico continues to be a colony of the United States.

Then came what is known as the Jones Act of 1917, where the American government granted U.S. citizenship to those born in Puerto Rico. To this day, as I write this book, the United States has complete control over Puerto Rico. During the First and Second World Wars, Puerto Ricans served in the US army. They served in Korea, Vietnam and any other place where American troops are sent. Military service was compulsory for every male who was at least 18 years of age, and that lasted until the 1960's. When I reached the age of 18, I reported to fulfill my duty as an American citizen. At that time, like it or not, that was the law.

During the Second World War, the First Puerto Rican Military Regiment was formally organized, which became known as the 65th Infantry Regiment. These men served honorably protecting the Panama Canal. This channel is the main route used by ships transiting from the Pacific to the Atlantic or vice-versa. The 65th Infantry Regiment went to the front lines in Europe and other places of military strategic importance to the U.S. Army or its Allies. When I was a child, I remember planes passing over my house to an unknown destination. They used power light reflectors looking for enemy planes in the black sky. During that time, the famous blackout came when least expected, which left the Island in total darkness. From my Uncle's house in San Juan,

I saw military aircraft trying to locate enemy ships and submarines approaching the coast of the Bay of San Juan. In 1945, I was seven years old. As a child, I used to listen to grown-ups discussing the trauma of war.

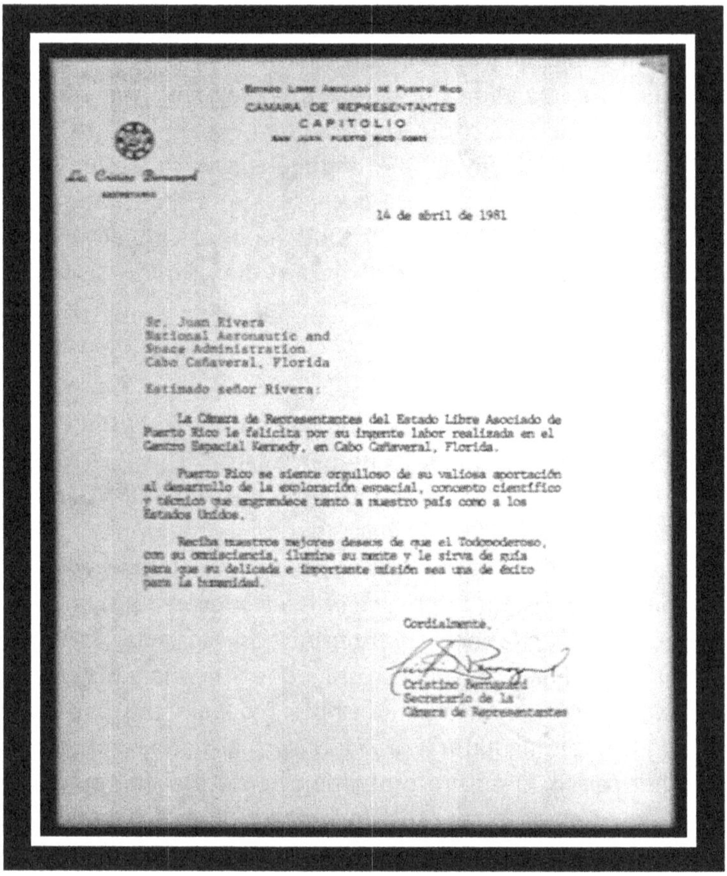

Estado Libre Asociado de Puerto Rico
CAMARA DE REPRESENTANTES
CAPITOLIO
San Juan, Puerto Rico

Lcdo. Cristino Bernazard
Secretario

14 de abril de 1981

Sr. Juan Rivera
National Aeronautic and
Space Administration
Cabo Cañaveral, Florida

Estimado señor Rivera:

La Cámara de Representantes del Estado Libre Asociado de Puerto Rico le felicita por su ingente labor realizada en el Centro Espacial Kennedy, en Cabo Cañaveral, Florida.

Puerto Rico se siente orgulloso de su valiosa aportación al desarrollo de la exploración espacial, concepto científico y técnico que engrandece tanto a nuestro país como a los Estados Unidos.

Reciba nuestros mejores deseos de que el Todopoderoso, con su omnisciencia, ilumine su mente y le sirva de guía para que su delicada e importante misión sea una de éxito para la humanidad.

Cordialmente,

Cristino Bernazard
Secretario de la
Cámara de Representantes

Figure 2: Letter of commendation.

My half-brother Manuel joined the Air Force as a volunteer. During those days, one had the choice of joining the military as a volunteer by meeting certain requirements. First, you had to be at least 18 years of age (although many boys, who were younger than 18, lied and joined. People used birth certificates of other family members with similar names. My cousin changed his name in order to meet that requirement); second, you had to have at least four years of education; third, you had to pass the physical exam; and last, you had to pass a written examination whose main objective was to separate the smart guys from the morons.

I noticed that the boys who served in the military provided a blessing for some families. It was said that one entered as a boy and came out as a brave man. The veterans of that era and their close relatives had many fringe benefits. For example, while my brother Manuel served in the Air Force, my mom received monthly assistance. With that money, she could buy clothes, shoes and food for me. Many soldiers' families received military aid as part of their service. Even the families of the soldiers who died on the battlefield also received benefits. Of course, one cannot measure the value of a loved one in dollars.

There is some confusion between the role of the Puerto Rico State Police and the National Guard. The Puerto Rico State Police has nothing to do with the National Guard or military service because they are two separate entities. One is military, and the other is civil in nature. In other words, the National Guard is not the police of Puerto Rico. But many who were part of the National Guard used it as a springboard to join the army. I think that one can still do it. The main purpose of the National Guard was, and still is, to help the people of Puerto Rico during natural disasters such as floods and storms. Another role is to ensure protection to the civilian population from an external force. The governor of Puerto Rico can call out the National Guard when the police or other agencies cannot cope with the situation at hand. The President of the United States may, at any time, order the National Guard to assist the U.S. Army to protect the nation. Because these mechanisms are there, the Puerto Rican National Guard attracts many candidates.

BRIEF HISTORY OF THE RADIO IN PUERTO RICO

I was extremely fortunate to have Don Genaro as my neighbor. As mentioned earlier, Don Genaro was a Dominican who fled the oppression of the dictator, Rafael Leonidas Trujillo. He had a series of short wave radios. His best radio was a multiband RCA VICTOR. It included the domestic AM band. It was a powerful radio that looked more like a piece of furniture than a radio. It had about 20 vacuum tubes, a 15-inch speaker, and sounded like a jukebox. He liked to raise the volume all the way up, blasting the immediate neighbors' houses. In the evenings, he listened to the shortwave band. We could hear many programs that originated in Havana, Cuba. Naturally, he was more interested in those stations whose programs originated from Santo Domingo.

He also listened to the Puerto Rican radio stations that transmitted their programs on the domestic band. At that time, there were three radio stations that could be heard well in my neighborhood. The first one was WKAQ in San Juan, then WNEL in Bayamon, and later we had WIAC with their studio in Santurce. There was also a station that I think belonged to the Department of Public Education because I could listen to programs such as those of Julia de Burgos and Enrique Laguerre, Lucy Boscana and James Washerman. The station was owned by the government and its sole purpose was to air educational programs. Of course, the dominant radio station was WKAQ that was founded in 1934. I remember listening to a program called "Los Embajadores Del Buen Humor" where the master artist was Ramon Ortiz Del Rivero ("Diplo"). There was another fun program hosted by "Rafael Agudo" that was known as "Los Jibaros de la Radio." WNEL began airing Sunday live music with Trios and Conjuntos. WIAC radio station aired their programs, but the signal was weak. I do not remember the date, but, by 1948, they had a station in Rio Piedras that they called WWWW, and the **cabin** announcer was "Pichin Roman." The cabin announcer was the guy in control of airing the program of the radio station. I enjoyed listening to WWWW because I liked the music from Cuba and they played mostly mambos of Perez Prado.

In 1949, Senator Ernesto Ramos Antonini presented a bill authorizing the creation of WIPR Radio as part of the Department of Public Education. The sole purpose of this facility was educational, airing programs created by local artists. The station would be the cultural heritage of the people of Puerto Rico. I sometimes listened to this because it was the only channel where one could hear the artistic works of famous artists such as Rafael Hernandez, Sylvia Rexach, Rafael Enrique Saldana, Abelardo Diaz Alfaro, Chuito el de Bayamon, Flavia Lugo, Marta Romero, and others. From the programming of that radio station were born many actors, musicians, singers, broadcasters, directors and screenwriters. The writers were the ones who laid out the foundation for what would become the basis of radio programming from 1950 until 1960. They were extremely talented people. As radio fell into the communications system, at some point in time, The Telephone Authority of Puerto Rico, which had some connection with WIPR radio, was created. At that time all public phones and the phones in many municipalities were government owned by this authority. The government of Munoz Marin created what was known as "authority." Let me give you some examples; The Port Authority, the Airport Authority, the Public Bus Transportation Authority and so on. I don't know what the state of affair of all of these authorities is. Those institutions did so much good to the island of Puerto Rico development. It will be foolish to sell these enterprises to privateers.

I have always been interested in electronics, especially radio and television. I was lucky that a boy like me, without any planning, absorbed the beginnings of radio in Puerto Rico and lived part of the time where this unique culture was being developed. Luckily, that had a strong influence later in my life. When electronics became one of my hobbies, I could better visualize the development of radio broadcasting not only in Puerto Rico, but also in the rest of the world. As this is a non-technical book, I cannot go deeper into technological discussions. So I have no other choice than to end the Puerto Rican radio report here.

INFULENCE OF BOXING IN PUERTO RICO IN MY TEENS YEARS

Speaking of radio, this brings me to the subject of boxing. All sports, including boxing, were sometimes aired by local radio stations. I liked listening to the fights on the radio. Many people spoke much about Joe Louis, who was one of my heroes. I was not especially adept at fighting, but I liked watching others engage in it. I remember listening to many Joe Louis fights on the radio. To me, he was invincible. When I started listening to his bouts, I became excited because he was excellent and had a lot of guts. The man had an outstanding record, never having lost a match. In 1936, he won the battle against the German Max Schmeling. In the southern United States, blacks were considered low class people, so it was inconceivable that a black fighter could beat a white boxer. The fact that Joe Louis won the match was also an insult to Adolf Hitler who believed in the superiority of the Aryan race. My family was not racist. To us, black and white were the same.

As a fan of Joe Louis, I continued following his career until his last fight. In my neighborhood, it was customary to meet a lot of boxing fans in front of the radio and listen to each round blow by blow. It is necessary to emphasize that the Island of Puerto Rico was and still is perhaps the only Caribbean island where boxing had always played a crucial role in our daily lives. I remember Joe Louis because he was a man who faced the best fighters of his era. At that time, there was much discussion about the following: Max Schmeling, Jack Roper, Arturo Godoy, Billy Conn, and Jersey Joe Walcott. The following heavyweight boxers exemplified the glory of the sport at the time: Tommy Burns, Jack Johnson, Jack Dempsey, Gene Tunney, Tom Gibbons, Primo Carnera, Ezzard Charles, Rocky Marciano and Floyd Patterson. Of course, there were other weight divisions that I will not discuss for lack of space.

When I was a kid, there was also a lot of talk about Sixto Escobar, who was the first Puerto Rican winner in the bantamweight division. He was born in Barceloneta and raised in San Juan. I still remember the old ball park bearing his name: the Sixto Escobar Ball Park near the Escambron Beach Club in San Juan. When I was a boy, I used to go there with my uncle. I think that ball park was torn down because it was a wooden structure. The heavyweight class division has always been dominated by people outside of Puerto Rico. The only fighter of Hispanic blood that I can remember was John Ruiz, who, in 2000, won in the heavyweight division. Without any doubt, Puerto Rico has had many talented boxers, and will continue to have them, as long as the promoters help them. As I used to live in Mexico, I should note that there are also many marvelous Mexican boxers whom I followed when I lived there. Boxing is like music. It has no frontiers or flags and has become truly global. The people that I mentioned here are the ones who came across my radar.

MY HIGH SCHOOL YEARS

I began high school in 1952 and graduated in 1956. I did not have any problems academically, and I never had to study because I had a photographic memory. My favorite subjects were algebra, plane geometry, solid geometry, trigonometry, physics, chemistry, biology and graphic arts. I did not enjoy the Spanish class because it was too tedious. We just had to copy down into our notebooks everything the teacher wrote on the blackboard. I enjoyed my English class, but I did not like the nun who taught the class. She was an extremely strict disciplinarian. I also had a terrible relationship with the nun who taught science and math. I never got decent grades in her classes because she thought I was copying from another student. She was wrong, but I figured it was better to ignore her than to fight a losing battle.

During high school, I was the Puerto Rican version of the rebel without a cause. I always felt that I was being discriminated against. It turned out that I was not the only one, so one day my friends and I formed a small group to get revenge on the nuns. Of course, the mischief we perpetrated was pretty typical of boys of our age. We turned off the lights, hid the chalk, hid the pencils, and threw paper airplanes.

Scholastically, I received many C's, a few B's and occasionally an "A." Being poor and having a rebellious attitude did not help. I did not care much about the grades because I knew that I was not a C student. I continued being an excellent listener and a skilled observer. What was on my side was that my father and my father's family were extremely wealthy people. They were Roman Catholics to the last drop. I also maintained close relations with the parish priests because they never harmed me. They knew my problems because, before receiving Holy Communion, they enjoyed listening to my confessions!

I have always had enormous respect for priests in general because, to me, any man who follows the rules of celibacy deserves a lot of respect. Opponents of Catholicism spent much time criticizing Catholic priests and nuns. Criticism has always existed and it always will. I never saw them perform any immoral acts with boys or girls. The nuns lived on the top floor of the school building, and it was impossible to climb to the top floor. First, the elevator buttons were not accessible to the fourth floor without a key; and second, the stairs leading to the fourth floor had padlocked iron gates. Only nuns and priests had keys. The nuns lived safely and protected.

I was always intrigued about what happened up there. As early as first grade, I learned that one must respect the security measures at the school. We never had access to the priests' living quarters, but nobody was interested in the priests' private lives. That was probably because we did not have daily relationships with them as with the nuns. I asked my mom why there were so many security measures, and her answer was always: "In a closed mouth nothing can enter." My high school years turned out to be extremely crucial to my future because I got a decent education, especially in mathematics and science, which just happened to be my favorite subjects.

MEDICINE AND WITCHCRAFT

From 1952 until 1954, I suffered a series of diseases that almost killed me. I had already survived diseases from gastrointestinal parasites. But more illness lay ahead. First measles, then chicken pox, and the rest of the childhood diseases followed. Months later, I suffered the worst disease in my life—diphtheria followed by extreme nausea.

For several months I have this nausea problem every morning while taking my breakfast. I didn't know what caused it neither the medical doctors, who prescribed me all kind of tonics and elixirs. Let me explain this situation in detail. In the morning, I sat down to have breakfast. My mom served me a bowl of oatmeal and a cup of coffee. Within five minutes, I felt nauseous, and I throw out everything that I ate. I had to wash my mouth and lie down for five or ten minutes to calm my stomach. But, the clock kept ticking and I have to eat something before going to school. So after a few minutes, I sat back down, ate whatever I could eat, then went to school and try to forget the nausea. At 10 a.m. there was a 10-minute break which gave me a chance to have coffee and maybe a small sandwich. I really was concerned because I was losing weight. At twelve, I ate a bit of lunch that was served in the school cafeteria. The nausea went away by magic. Classes ended at 3 p.m. and everyone went home. Instead, I went to the principal's office to help. Around 3:30 p.m. I walked home. Parents came to pick up their kids in brand new cars. Others who lived near the school walked home, but I had to walk 30 minutes to get home. I had to cross a stream and then a farm to get to my house. If it rained, then I had to cope with floods or else I had to go by way of a more distant road. I have never been afraid of anything and did what I had to do to get home. There was times that I to confronted bully children that enjoyed picking on other kids like me. At other times, I have to face bad dogs. But, the scariest situation for me was to face the MALOTES. Yes, those bad guys that were looking for a victim to get money to pay for their vices. I just walk past them without looking at them. Fortunately they never touch me. I could hear their comments among themselves whispering; "That is the son of Pedro 19." People called my father "Pedro 19" because he has 19 fingers. He lost his right hand thumb in an accident with a machete. My father use to give money away to the MALOTES to keep them happy. My father fears them too. So my father was in the MALOTES list of victims.

Finally I could see my house. My house was my kingdom. There, my mom treated me like a king. After I bathed, I would listen to the radio. Sometimes my friends came and then I asked my mom for permission to let me go and play with my friends. Since my neighborhood had a lousy reputation, my mom worried that something bad might happen to me. She was afraid that my father would come and take me away. Fortunately I never had a problem with my friends. So I told my mother not to worry because I was happy with her and my half brothers and sister. It never crossed in my mind the need to go back to my father.

My mom took me to the best doctors in town. Dr. Ramirez-Santos and Dr. Garrido-Collazo knew me inside and out. At that time, the best doctors were really the pharmacists! There were three pharmacists who already knew me because I had to go to them almost weekly. My mom went to the pharmacists and explained my health problems to them. They immediately prepared some tonic which I drank according to instructions. Some of those tonics worked; others merely put me in a mental trance. I felt that I was floating between heaven and earth. One day I read the label, noting that it said "contains 20% alcohol." No wonder I felt like I wanted to dance. I told my mom that I did not want to drink any more of that alcoholic stuff.

Of course, my mom was extremely worried about my health and was forced to sell her old fridge to pay for my medical expenses and medicines. She decided to take me to a lady who was called "the witch." She was considered a healer through her use of herbs. She believed in the African religion of "Yoruba." She was a kind of sage or priest (a "babalao"). She placed me in a large metal bucket full of unknown herb leaves which were exceedingly fragrant and intoxicating. Then she cried out to my mom, "this child was cursed by a malicious witch, and you need to get the sickness out." My mom asked her how much the procedure would cost. The witch replied, "It costs $30, but if you do not have the money, you can pay me in installments." The lady then made me drink a strange juice that tasted worse than castor oil. I vomited green stuff so violently that I feared my guts would come out too. But in a few days, I was eating normally. My mom told me that maybe "someone cast an evil eye on my baby." I am not superstitious, but I believe that some people envied me and probably wanted to see me dead. In Puerto Rico, as well as in other islands and other countries, there are persons who engage in sorcery. In my neighborhood, they called them "spiritualists." I believe the common root is the Yoruba religion that was brought over from Africa. There are many superstitions in the Island and many people practice variations of it. Even in my family, people practice these things behind their priest's back.

During my junior and senior years, I was finally clear of stomach problems. Just to be safe, I followed the advice of pharmacists of that time. Alcohol was an extremely beneficial ingredient in many medicines. So I believed that a cup of wine before a large meal helped the digestive tract. The important thing to remember is that like everything in life, "excesses are always bad." Therefore, I suggest being extremely careful with the ingestion of any alcoholic beverage. My dad and some of my family used to drink every day. Many of them paid with their lives for not knowing how to control their intake of alcohol. It is a sad fact that many of the best artists and a number of professional people, including lawyers, doctors, writers, painters, and also people of lower classes have become victims of alcohol. My father used to say: "if you play with fire, you will get burned." I have sometimes had bottles of liquor of different brands and flavors set temptingly in front of me, but I do not dare touch them because I know the consequences.

MY FIRST JOB

When I turned 15 in 1953, my mom convinced me that working could be particularly beneficial. Money earned should be spent wisely, and part of it should always be saved. She talked to one of my cousins who owned a small grocery store about hiring me as a helper.

On my first day at work, my cousin Johnny taught me how to cut a tail off fresh dried cod fish. At that time, they used the bottom of the trunk of a tree as a cutting table. I learned to cut cod fish and other types of meat with a sharp or hatched knife which was sharpened like a razor blade. Because someone was paying me for my work, I learned the value of a dollar. He only paid me half a dollar a day, and I did not like that. I told my mom that I wanted to earn a dollar a day, but Johnny told us that legally I could not work because I was a minor. So I had to wait until I turned 16 when, with the written permission of my mom, I would be able to work. So in the long run, Johnny helped me.

I had two cousins. One was the owner of "La Rivera," where my mother worked as a seamstress, and the other was the owner of "La Colombina" in the center of Rio Piedras. Later on, one of my cousins opened another store called "La California." I began to work there on Saturdays when I was 16 years old and when I was not in school. I continued working there until I turned 18. My cousin paid me a dollar a day. To me that was a fortune as a coke cost a nickel and a delicious sandwich cost a quarter. Movies cost 15 to 20 cents. I saved my money so I could go to the movies on Sunday afternoon.

Working in that store gave me a solid education because I learned (1) to deal with people; (2) to use mathematics; (3) to make suggestions to my supervisor (my cousin), and (4) to understand the value of a dollar. One day, I went to get a pair of shoes for a customer and found a stack of pants with worms. I went to my cousin and asked him what he planned to do with the pants because he was going to lose them all. He asked me, "What do you want me to do?" I could see that he was upset. I replied, "Well, I would put all those pants outside on the sidewalk and put up a large sign that says "ON SALE." I told him to add in small print: "some pants have flaws; no refunds." In a few hours, all the pants were sold out. He rewarded me with a new five dollar bill! That was the first instance that revealed my aptitude for business.

One day a lady asked to see a pair of shoes that she had seen in the window. I went to the window and saw the shoes, but when I went to the warehouse, I had forgotten the style of the shoe and ended up bringing her the wrong shoes. The lady was furious and insulted me. I sat down looking for a solution to that problem. It was then that I went to my cousin with another idea. I suggested that he put a unique code on the box of every shoe, and then put the same code inside the shoe with a sticky label. That way, if a customer asked me for a pair of shoes, all I had to do was look inside the shoe for the code, and then when I went to the warehouse, I would know exactly where the boxes corresponding to the style of that shoe were. Every style had a unique number. He liked the idea and implemented it the same week. When the next customer approached me and pointed to a particular style, I just had to look at the code number in the shoe, and the rest was easy. My idea was a smashing success. My cousin was so pleased he gave me a new ten dollar bill!

Other shops copied the idea, and it spread like wildfire. Those made me realize that five or ten dollars was a pretty paltry sum for coming up with a good idea. I decided then and there that I would not give any more ideas if they were only going to pay me a pittance. So despite my initial joy in getting the ten dollars, I realized that I had not been given any credit or benefited in any way from my idea. I was young and did not know that there was a patent office in Washington DC. I learned about patent law many years later. The other lesson I learned was that I had absolutely no future as a store salesman. I could not envision spending the rest of my life selling shoes. My cousin had employees who had worked for him for more than twenty years, and they were still living in rented houses and had no savings accounts in the bank. A man's life totally wasted. Meanwhile, my cousins were getting richer and richer. So I learned my lesson quite fast.

One day I went to my mom and said, "When I finish high school, I want to study at the University of Puerto Rico. I want to become an engineer or a medical doctor. I do not want to spend the rest of my life selling trinkets in a department store." My mother looked at me with tears in her eyes and said, "I would do everything possible to get the money to send you to school." Finally, in May of 1956, despite all my troubles through high school, I received my high school diploma with joy and pride. My foster mother, who sacrificed so much to help me achieve that goal, was so proud of me. To this day, I still have my high school diploma on the wall in my private office.

UNIVERSITY OF PUERTO RICO BACHELOR DEGREE

During these four years, many fascinating events happened in my life. After graduating from high school, I asked my mom if she could pay my tuition for the first year at the University of Puerto Rico, Rio Piedras campus, better known by the acronym UPR. I promised my mom that after a year, I would cover my own school expenses by working odd jobs. I had been admitted to the UPR Mayaguez campus (CAAM) which was an engineering school, but it was too far away from where I lived. The UPR main campus was within walking distance from my house. Plus, it was cheaper to attend the UPR Rio Piedras campus.

My first year at UPR was quite difficult because I had to adjust to another way of life. I knew that to be successful, I would have to hit the books. My student average was a flat B or 3.0. With that index and being considered a poor kid, it placed me in the upper quartile of incoming freshman students. That meant that I was going to be able to get the best teachers. My English teacher was an American lady who was remarkably talented and, as she did not know any Spanish, only spoke in English. That was no problem for me because I learned to speak English as a child. The Spanish teacher was a remarkably intelligent and aristocratic lady. We called her Professor Ramos-Mimoso. She was an excellent teacher, but on the first test, I got an "F." That failure was a monumental embarrassment for me. I took the exam, wrote a large "F" on the paper and stuck it on the door. That day, facing the enormous "F" nailed to the door, I swore to never get another "F." Daily, I put all my effort into studying extremely hard for that Spanish class. Then I took the second test. I was surprised when I saw the score—an "A!" So averaging the two grades, my overall grade in that class went to a C. I never removed the first "F" from my door. I left it there for a year to remind myself that I could get an "F" in that or any other class. My humanities teacher was Professor Oliveros. He was a highly intelligent man who also came from a poor family. He was a self-made man whose success in life inspired me. In that class, I received straight A's. Another excellent teacher was Professor Arsenio-Torres, who taught social sciences. My physical sciences teacher was also the head of the Physical Sciences Department. He was a Jewish expatriate who had fled from Hitler's Germany. He spoke with a German accent mixed with English. Although he understood Spanish, he was not truly fluent conversing in Spanish, so he taught the class totally in English. I took college algebra with the head of the Mathematics Department. He was getting up in years, but he was still an excellent mathematician. Besides profoundly inspiring my career as a mathematician, Professor Laborde was also my role model in teaching me self-discipline. My botany professor was a woman who was exceptionally knowledgeable in her field. My zoology teacher was Professor Pesquera. He was a brilliant man who also taught me how to be an organized person. Because I was skillful in dissecting animals, Professor Pesquera wanted me to study medicine, but I told him that I preferred mathematics and physics.

My freshman year was a difficult one for me. I lost 10 pounds and dropped to 110 pounds. It was customary in that school environment for classmates to give everybody a nickname. I was called the "skinny Socratito. A year later they called me "Galileito."

At the end of my freshman year, my school average rose from 3.0 to 3.5 and that made me eligible for a Puerto Rican government scholarship. My scholarship paid all my school costs including tuition and books. Therefore, I never had to pay a penny for my education. They even gave me a stipend to cover expenses such as food and transportation. The stipend was in the form of a monthly check. That meant that I did not have to work at odd jobs to pay for my education. My only job was getting A's or B's. I used my leg muscles instead of paying for transportation, so I saved that money in a piggy bank. It was vigorous exercise to walk daily from my home to school. I was skinny, but it was all muscle. I could have used my bike, but I was afraid that somebody would steal my "Schwin Majestic" that was my treasure. I went to the library to study during the day to avoid the walk home.

In my spare time, I became hooked on building model airplanes. I used to fly them on Saturday at the University athletic field. After a while, I felt like a jerk just flying them around in circles using a wire line. Then I became interested in building radio-controlled airplanes. We used vacuum tubes in the airplane receiver and vacuum tubes in the airplane transmitter that you held in your hands. Unfortunately, I crashed several of those radio-controlled airplanes and lost my money, so I decided to quit. The time was the late fifties when electronics were based on vacuum tube technology. I sold all my airplanes and gave whatever was left to one of my cousins.

My government-sponsored scholarship was a gift that came to me when least expected. I took that opportunity to get a solid education. I became an A student in mathematics and physics. As a college student, I had access to medical services that UPR offered to its students. That was crucial to me because if I got sick my mother would not have to worry.

In today's world, I am truly sorry to see how bankers and other financial institutions are exploiting the student population. In other countries of the world, education is free, even at the University level. Other countries charge an unusually low fee to get a proper education. In the United States, one has to nearly sacrifice his future to get a university degree. Of course, the same resources that I had are still available, but they are more difficult to obtain and seem to be reserved for a chosen few.

I used to go to see my previous coworkers at La California store to tell them that I was getting a University degree. They applauded my efforts without displaying any envy or bad will. After I discontinued my airplane hobby, I became a movie fan. I went to the movies every Saturday night and Sunday afternoon. I watched different movies—

Mexican, Cuban, Argentinean and of course Hollywood cowboy movies, especially Roy Rogers, Gene Autry, Zorro, etc. I never missed a Tarzan movie with Johnny Weissmuller. He was my favorite movie actor. More than half a century later, I am still a film buff.

From UPR to my house was a 30-minute walk. I still remember the names of those streets because I spent so many days of my life walking them, rain or shine. At that time in Puerto Rico, there was a passenger train that ran behind the UPR campus. There was also a train station, opposite the rear UPR exit. The railroad was the backbone of the sugar cane industry. The passenger train enterprise was a real failure. For some reason, the Munoz Marin government began dismantling the railroad industry. I noticed that every year, the rails were pulled out from the ground and the ground was left bare. People used it as a pathway or road. I used to walk every day through the neighborhood of Capetillo. I had to pass in front of Chaguin's Bar.

I could hear the excitement and music of trios coming out of the doors. One day, I got up the nerve to enter, and what I saw took me by surprise. Felipe Rodriguez, Pedrito Berrios, Henry Vasquez, Tito Lara, and other recording artists were there. They looked at me with my books and laughed. One of them shouted, "Boy, what are you doing here?" I answered without hesitation, "I am a college student, but I also love good music and just want to sit at the bar and have a beer." Then I walked to the bar, ordered a beer and listened to the fantastic songs sung by those people. What struck me the most was that they started telling funny stories and jokes. At that time in Puerto Rico, there were no laws that prohibited serving drinks to minors. I started drinking beer in public when I was fifteen. The only problem was having the money to buy beer. Bars never served rum to a minor because rum was more controlled by the government. One could bathe in beer, and nobody would say anything. That night, I only had money for two beers, and I needed to stretch it out for several hours. From that day on, I went by myself almost every Friday, had a beer, and listened to the music and the stories of those artists. They must have kept that place a secret because it was in a secluded spot and few people knew that it existed.

From Monday to Friday, I dedicated 100 percent of my time to the courses I was taking. Sunday morning was devoted to my Catholic faith. I enjoyed going to church with my mom or my friends. My family was passionate about their Catholic faith. I was not, nor have ever been fanatic about any religion. I practice my religion and respect other religions. I also understand those who call themselves atheists, and those who go to church two or three times a year. I mind my own business. I expect people to respect my beliefs.

ASSISTANT PROFESSOR AT THE UPR

By this time I had a girlfriend, but due to conflicts between my girlfriend and my mom, the relationship suffered a lot. The biggest problem was in the matter of religion. I was Catholic, and her family was Protestant. To make matters worse, my girlfriend's father was a member of the Masonic Lodge. For some reason, the Catholic Church was not sympathetic to Masons. To complicate things further, my foster mom did not allow me to bring my girlfriend to the house. I was getting ready to break those stupid barriers because, for me, the decisive thing was the love between two persons, not religion. My Catholic family was not happy and gave me a hard time. Between the religious mismatch and my family's criticism, it was inevitable that my girlfriend and I would break up.

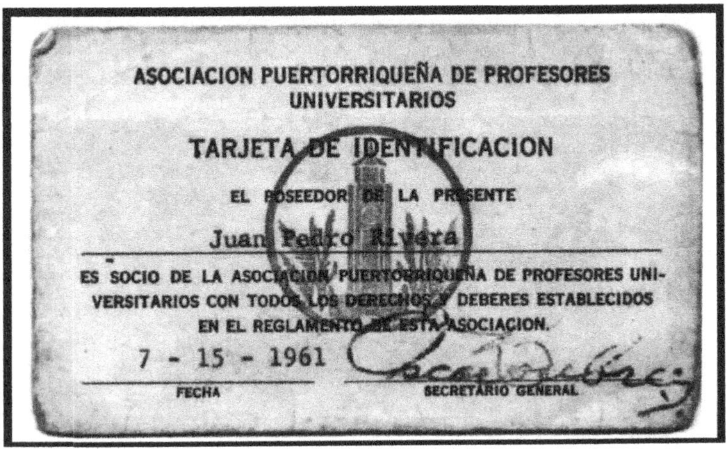
ASOCIACION PUERTORRIQUEÑA DE PROFESORES UNIVERSITARIOS

TARJETA DE IDENTIFICACION

EL POSEEDOR DE LA PRESENTE

Juan Pedro Rivera

ES SOCIO DE LA ASOCIACION PUERTORRIQUEÑA DE PROFESORES UNIVERSITARIOS CON TODOS LOS DERECHOS Y DEBERES ESTABLECIDOS EN EL REGLAMENTO DE ESTA ASOCIACION.

7 - 15 - 1961

FECHA

SECRETARIO GENERAL

Figure 3: Association of Puerto Rican Professor ID card

I maintained close relations with her parents, her sister and her family. Tragically, my girlfriend's brother was killed in a plane crash along with his wife. Her older brother was studying to be a dentist, and we used to talk a lot about different subjects. My girlfriend had an uncle who was also a close friend of mine. Every Friday I went to his place of business to socialize with friends. It was there that I met the famed Puerto Rican composer Rafael Hernández. I sat at a table talking to him about music and composers. There was another famous composer, Edmundo Disdier, who sometimes brought his guitar, and we had a ball singing and serenading. When my girlfriend's older brother and his wife perished in a plane crash in the Bermuda Triangle, the whole family went into mourning. After I broke up with my girlfriend, I decided to commit myself to my "alma mater"—UPR.

Returning to academics, in my second year at UPR, I decided to declare mathematics as my major with physics as my minor. By now, the mathematics teachers knew I was proficient in mathematics—so much so that I completed my math major requirements in three years. During my junior year, Dr. Pinero, one of my mathematics professors, called me to his office. He asked me if I wanted to earn some money. I asked him to describe what my duties would entail. He told me he needed me to be his assistant. My main responsibility was to grade students' homework and exams. Later on, I began tutoring his students. That job was a piece of cake.

The UPR chancellor, Don Jaime Benitez, instituted a program that became known as "Extramuros." This word means "outside the UPR walls." At that time, there were only two campuses of the UPR university system—the Mayagüez campus and the Rio Piedras campus. The Rio Piedras campus was the main campus. Don Jaime saw the need for UPR to branch out to other key cities. This would give many public education instructors an opportunity to be more proficient in mathematics and science without having to travel to Rio Piedras. Every Saturday, extramural teachers would teach in those cities from 9 a.m. until 2 p.m. Using the facilities of the local public high schools, they chose a school that had exceptional physical facilities and was located in the center of town. The centers were located in Ponce, Arecibo, Vega Baja and Humacao. The concept was an enormous success and gave rise to what became Regional Colleges. These Regional Colleges offered the first two years of courses called General Studies, which are the courses that all high school students have to take before focusing on their majors. At the end of the two years, students could continue studying in Rio Piedras or Mayaguez. The idea was to decongest the UPR campus in Rio Piedras by making better use of existing physical facilities available at that time. Then the UPR chancellor started building new facilities in Rio Piedras and expanding the land area used for education. This idea was not original. In the U.S., the same model is called "Community College." But Don Jaime improved the idea to make it adaptable to the requirements of Puerto Rico. I believe that Puerto Rico owes him an enormous debt of gratitude.

The information that we gathered from statistical analysis showed that out of 100 students entering the first year of college, by the end of the second year, there might be 40 students left. This loss is called attrition. The Regional Colleges became the sieve or filter. The result was that we were able to make better use of the physical facilities of the Rio Piedras campus. I used to do the statistical analysis for my professor who, in turn, was involved in those studies.

I started to work as an assistant to Dr. Piñero in 1958. I still have the contract that was signed. Please refer to figure 3A. One day I said to Dr. Piñero that by the end of my third year I wanted to teach at the Rio Piedras campus. When I finished my third year and had completed all requirements for my mathematics major, I became a physical sciences instructor. That happened in 1959. I just taught one class because I was also finishing my physics requirements in my senior year. In May 1960, I graduated from UPR with a Bachelor of Science degree with two concentrations, one in mathematics and one in physics. My grade point average for the four years went up to 3.6, and I graduated *Magna Cum Laude*.

To my immense surprise, the head of the Physical Sciences Department at that time was Prof. Rivera (no relation). One day he called me into his office and asked me if I wanted to teach physical sciences. Without hesitation, I replied, "Yes!" At the end of my senior year, at the young age of 22, I started my career as a full-fledged professor!

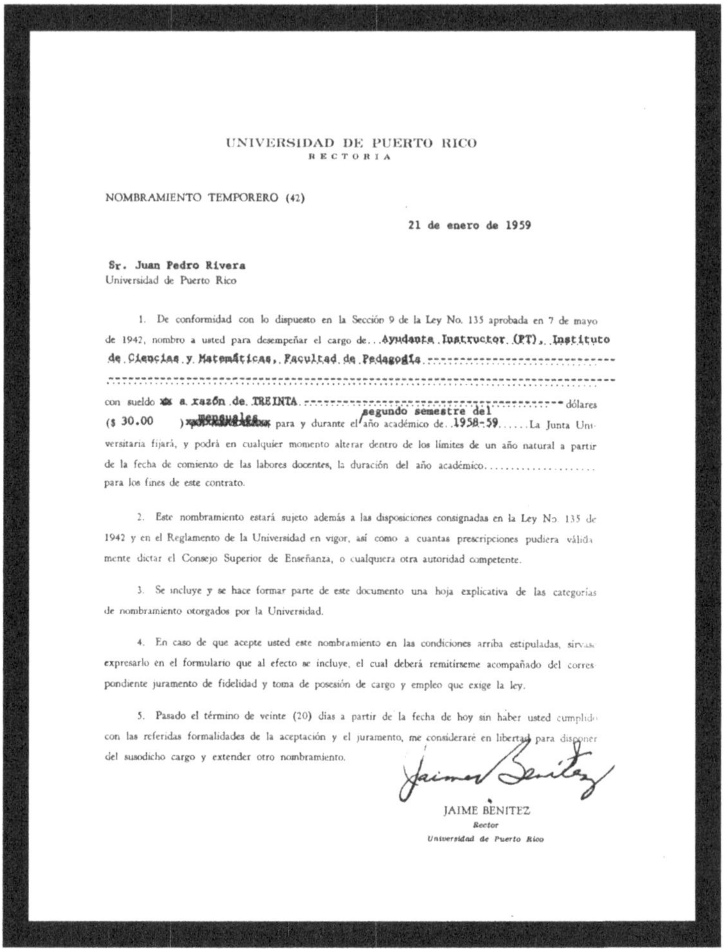

UNIVERSIDAD DE PUERTO RICO
RECTORIA

NOMBRAMIENTO TEMPORERO (42)

21 de enero de 1959

Sr. Juan Pedro Rivera
Universidad de Puerto Rico

1. De conformidad con lo dispuesto en la Sección 9 de la Ley No. 135 aprobada en 7 de mayo de 1942, nombro a usted para desempeñar el cargo de Ayudante Instructor (PT), Instituto de Ciencias y Matemáticas, Facultad de Pedagogía

con sueldo a razón de TREINTA dólares ($ 30.00) mensuales para y durante el segundo semestre del año académico de 1958-59. La Junta Universitaria fijará, y podrá en cualquier momento alterar dentro de los límites de un año natural a partir de la fecha de comienzo de las labores docentes, la duración del año académico para los fines de este contrato.

2. Este nombramiento estará sujeto además a las disposiciones consignadas en la Ley No. 135 de 1942 y en el Reglamento de la Universidad en vigor, así como a cuantas prescripciones pudiera válidamente dictar el Consejo Superior de Enseñanza, o cualquiera otra autoridad competente.

3. Se incluye y se hace formar parte de este documento una hoja explicativa de las categorías de nombramiento otorgados por la Universidad.

4. En caso de que acepte usted este nombramiento en las condiciones arriba estipuladas, sírvase expresarlo en el formulario que al efecto se incluye, el cual deberá remitírseme acompañado del correspondiente juramento de fidelidad y toma de posesión de cargo y empleo que exige la ley.

5. Pasado el término de veinte (20) días a partir de la fecha de hoy sin haber usted cumplido con las referidas formalidades de la aceptación y el juramento, me consideraré en libertad para disponer del susodicho cargo y extender otro nombramiento.

Jaime Benitez

JAIME BENITEZ
Rector
Universidad de Puerto Rico

Figure 3A: My first Professional Professor Contract

At that time, one had to teach four groups or what they called physical science groups. That was considered a full academic load for a college professor. I must point out, for those who do not know, that in addition to the work done in the classroom, a teacher has other responsibilities. Sometimes those responsibilities take more time than that used to prepare lessons. Outside the classroom, one had to grade papers, grade exams, and still have a few hours to meet with students. Sometimes I spent more time in the office than in the classroom. On top of that, a professor must go to meetings, symposia, read, read, and then read some more. It is a time-consuming job. Then one day you realize that your salary in academia is low compared to jobs outside the classroom. Do not get me wrong. Teaching is an extremely fascinating job. No two days are the same. I loved it.

How proud my foster mom and my foster brothers were! My real parents, brothers and sisters, were also happy to see me achieve my goal. How could anyone ever imagine that that skinny, scrawny boy was a genius in mathematics and physics? To me, it just came naturally. In short, from 1956 to 1960, my life made an incredible transformation that catapulted me onto another level. My first salary was $300 a month. I used that money for the down payment of a car that cost over $1,000 with interest and other charges. Everyone remembers his first car. Mine was a 1953 Chevrolet Belair four-door sedan with a four-speed manual transmission. I was thrilled with my first car. With my new mode of transportation, I no longer had to walk considerable distances. I remember people in my neighborhood scratching their heads to see a 22-year-old boy driving his own car. They could not believe it. People who saw me grow up in my neighborhood were proud of me. It meant that one of them could also achieve success early in life.

I would like to take this opportunity to pay tribute to my college professors at UPR at that time. The ones I remember are my teachers of physics: Rafael Arce, Santiago Ramirez, Dr. Leticia del Rosario, Dr. Amador Cobas, Prof. Felix Pereira, and Prof. Osvaldo Matos, who did so much for me; my math teachers: Prof. Enrique Bayo, Dr. Pedro Laborde, Dr. Alfredo Piñero, Prof. Jesús Lima, Prof. Jesús Bajo and others whose names I no longer remember. I also want to pay tribute to some of the professors of other subjects who were also extremely influential in bringing me to the academic level that I achieved: Prof. Antonio Oliveros (humanities), Prof. Arsenio Torres (social sciences), Prof. Kiko Pesquera (zoology), Prof. Facundo Bueso (physics), Dr. Roberto Diaz (mathematics); Deans and Heads of Department: Ken Manfred, Prof. Pedro José Rivera, Don Jaime Benitez (chancellor), Dr. Daniel Curet (Dean) and others. Many of them have passed away, but I want to express to their families how instrumental all these people were in drawing out the hidden genius that was always inside me. In my opinion, they were the real heroes.

PROFESSOR OF PHYSICAL SCIENCE AT UPR

Don Domingo Marrero was the Dean of the College of General Studies and, to some extent, was my first boss. I was a member of the Faculty of Physical Sciences at the College of General Studies, which was the name Don Jaime Benitez preferred when referring to this faculty. Dr. Facundo Bueso was the Dean of the College of Natural Sciences who was one of my teachers in physics. Although Don Domingo was relatively young, perhaps about 40, I noticed that he had difficulty climbing the stairs from the first floor to the third floor of the building. Months later he fell ill and had to resign his position. Dr. Ethel Ríos de Betancourt filled in for him and Dr. Quintero Alfaro was her assistant dean. One day she called me into her office. She asked me if I was interested in getting my Ph.D. I told her that I preferred to teach a couple of years before going to graduate school.

Prof. Pedro José Rivera was appointed Chairman of the Physical Sciences Department. My name on the payroll was Pedro Juan Rivera, and my check came payable to Pedro J. Rivera. Checks for Pedro José Rivera also came under the name of Pedro J. Rivera. One day, I opened the envelope, took out my paycheck, and found that they had raised my salary. For a short time, I was happy. The next day, Pedro José Rivera called me into his office and told me he had received my meager paycheck instead of his juicy check. Then he said, "It is better we put you on the payroll as Juan Pedro Rivera instead of Pedro Juan Rivera." To this day, I've remained Juan Pedro Rivera. I replied that it was no problem as long as I retained my position as a professor of physical sciences. On all school records, my name is Juan Pedro Rivera, although my birth certificate says otherwise. I never let anybody mess up my name again.

Another of my activities at the University of Puerto Rico in the 1960s was as followed. When I was working in

"Extramuro" which means beyond the walls, I became a friend with two professors who worked on EXTRAMURO. I called them the Melendez brothers. At that time, I was extremely familiar with Professor Ramon (Moncho) Melendez who was intensely interested in the formation of an association of the faculty. We also talked about other issues such as the insular and foreign policies. There were some issues of concern that we discussed each time we met. One was the political conflict in Cuba under the leadership of Fidel Castro. I was pretty soaked in what was happening in Cuba before and after Fidel Castro. Another concern was the need of some sort of union or association to protect the rights of the faculty at the institution. Sometime around 1960, the "PUERTO RICAN ASSOCIATION OF UNIVERSITY PROFESSORS" was formed. I became one of its members. I consider myself one of the founding members of this noble association. APPU, the acronym used to refer to this association was growing in the number of membership. Please refer to Figure 3 showing my identification card APPU. I was extremely pleased to join this group on July 15, 1961. The Melendez brothers were the dynamo that kept the alliance moving forward.

We were able to obtain official recognition from Don Jaime Benitez. It became a group of professors working together in a partnership to defend their rights. At that time, the administration provided a building that was the meeting place of the faculty. It was called the Faculty Club. We took the opportunity to gather there to exchange ideas. The building was later used by the Faculty of Architecture. Well, in that building we have a convention center for the exclusive use of the faculty. We had several rooms with ping pong tables, a bar and grill for cooking, and many other types of entertainment. There were rooms that were used to have faculty meetings. Later, we added some pool tables. A committee was formed to help faculty members to serve as arbitrators between the administration and any faculty member who needed help. From what I can remember, one of the main themes was to help faculty to obtain their permanent job status. There were different pay scales, from what I remember the pay scale was; Assistant Professor, Associate Professor and Professor. There was some arbitrariness in getting those positions. The UPR was becoming politicized.

They informed me that one day facilities were closed because they were going to use it for the new School of Architecture. We felt extremely unfortunate because we lost a gathering place. I told Moncho that UPPA will never die. It was just a matter of finding another venue. I continued as a member, and I supported the organization. When I left the island to work for NASA, I lost touch with this important group of faculty members. In my opinion the greatest achievement was that we planted the seed. The group continued to grow and is a big organization nowadays, thanks to the vision of the Melendez brothers and other teachers who supported them. I am glad that the APPU remain functioning and growing. Unfortunately, I am no longer a member of that noble association. Today I am retired from the University of Puerto Rico and receive a small pension. I will make an effort to reestablish a connection with APPU.

I liked teaching the course of physical sciences because it was particularly appealing. Here, I was at age 22, a professor of the UPR system with all the benefits and prestige that the position carried. I had already bought my first car, and my mom told me that it was time for me to buy some property. My sister Toña had moved to a subdivision called Country Club and the properties at that time were within the reach of my pocket. So one day I went and bought property in the second extension of that subdivision which is in the jurisdiction of San Juan. My house was near the house of my half-sister Toña, whom I loved and who was an influential person in my life. She was the cook of the house while my foster mom worked as a seamstress at La Rivera. I paid the down payment for the property which was only $300, and the monthly mortgage payments were only $63.00. The house payments made no significant dent in my monthly salary. With the money that I earned, I had enough to pay for the house and my car with money left over for other necessities of life. On top of that, the Dean gave me an opportunity to teach on Saturdays in EXTRAMURO. Dr. Pedro Laborde who was the Mathematics Department Chair assigned me a mathematics course to teach at night. I had three checks coming into my pocket, and my entire academic load was easy to deal with because I was in my prime years. Thank God that my health was excellent.

STUDENTS PROTEST IN THE UPR 1960

At that time, UPR was a political hotbed of both extreme right and extreme left. There were two events that politicized the University and the people of Puerto Rico. First, both Alaska and Hawaii were admitted as states of the American union. This meant that to be state of the United States, it was not necessary that the territory be contiguous to the other 48 states as argued by the enemies of statehood for Puerto Rico. Second, the Pro-independence University Students Federation, better known as FUPI, began a campaign to abolish ROTC on campus. The ROTC (Reserve Officers Training Corps) was offered at UPR at the time and attracted many students. This is a program offered in thousands of U.S. colleges around the USA. Those who enter this program had their college expenses paid by the U.S. Military.

Upon completion of four years of study, the cadet earns an academic degree and also becomes an officer in the U.S. Army. One requirement is that after graduation, the cadet is required to serve four years in the U.S. Army. Male students had two options at that time. One was to take two years of military drill without having to enter the cadet program. The second was to take the complete ROTC program. Some people argued that they certainly were forcing students to play the soldiers game. This was not acceptable to pro-independence students. Students who, for medical reasons, could not stand under the sun had the option of taking two years of gymnastics. When I was a student, I chose to take the Gymnastics program instead of ROTC by using a medical excuse. I spent two years taking exercises leading to the famous Intercollegiate Tournament. Prof. Eugenio Guerra was my teacher, and his assistant was athletic Prof. Fuster. As I weighed only 110 pounds, I was exceptionally skillful at running the 100 meters and 400 meters, with or without obstacles. Students used to say that I was the "running thermometer."

Due to the student revolt and the pressures on being a teacher, I had to decide to defend either statehood or independence. Some friends convinced me to help them organize what became known as AUPE, which was a pro-statehood group. So without thinking of the consequences, I became a founding member of AUPE. I believed at that time that it was better for Puerto Rico to become a state of the Union. In that effort, I met the leaders of the Republican Party of Puerto Rico--Don Luis A. Ferre and Don Miguel Angel Garcia Mendez. Later on, I met Don Ramiro Colon who had plenty of money to fill the party's coffers. I remember working closely with students such as Georgie Navarro Sr., Edison Misla Aldarondo, and Gilberto Cruz. We were actually at odds with the opposite group that defended the idea of independence.

Later I was elected President of the Youth for Statehood group under the umbrella of the Republican Party for the city of Rio Piedras. I organized the youth to have representation in every neighborhood of Rio Piedras. I became real close with leaders of that time: Don Miguel Angel Garcia Mendez, Don Luis A. Ferre, and Don Ramiro Colon. Someone apparently blew the whistle on me to my supervisor at UPR. One day, the dean called me and said he was sending me to study for my master's degree. They offered me a scholarship which was essentially a form of financial aid. The institution would pay all of my education expenses, i.e., tuition, books, housing, meals and airfare. I did not hesitate; I simply answered with a big "YES." I had already been accepted to two universities. The first one was New York University, and the second one was Columbia University. After studying the pros and cons, I decided to study at Columbia University because Columbia University was, and still is, a prestigious institution of higher learning.

I had a meeting with my assistant in politics who, at the time, was a student, and was also my helper. His name was Edison Misla Aldarondo. I explained my future plans to him and told him that I supported him to replace me as the president of the Youth Group within the Republican Party. Both George Navarro Sr. and Edison Misla Aldarondo were friends of mine, and I told them it was more beneficial for me to get my Master's Degree than to play the political game. Politics is certainly a gamble. One may win, or one may lose. If one wins it is OK, but, if one loses, then one is in significant trouble. Having a graduate degree is like having money in the bank—you can lose your money, but you will never lose your degrees. My two friends communicated my plans to those who might be concerned, and they said that they understood my decision. They told me that the door was always open for me to come back. Looking back, I believe that I did the right thing. Being in the political game is like being on a roller coaster, and I had decided to get off. In the summer of 1963, I left for New York to begin my graduate studies.

MY MASTER OF SCIENCE AT COLUMBIA UNIVERSITY

The change from living in Puerto Rico to living in New York was extremely difficult for me. It was the first time I lived by myself. It was also the first time that I ate hamburgers and hot dogs and felt the seasonal changes in the weather. I began to study electronics. At that time, all electronics were based on vacuum tubes, resistors, capacitors and inductors. In the summer, I took a laboratory course in electronics because it was one of the requirements. My first contact with a teacher was Professor Lucy Hayner, who was an excellent laboratory teacher. She was about 50 and had a solid foundation in electronics. She taught all laboratories of physics, but electronics were her field of expertise. As I was only 25 years old, it was easy for me to learn all that technology because I studied really hard for that class. I also learned to read schematic drawings and electronic symbols. The field of electronics has always been one of my hobbies. I was hugely successful in that course, earning an "A." The next step was to enroll for courses that would start in August of 1963.

Days later I woke up with a sore tooth, but I did not pay too much attention until my face swelled the following day. I was living at the school dormitories, and they took me to the emergency room at Lincoln Hospital. I was told that I had two infected teeth which had to be pulled out. They were the upper two front teeth. I agreed to the surgery. The anesthesia that they used was commonly known as "laughing gas," which made me laugh continuously for about 30 minutes after I woke up. When I looked in a mirror, I saw that I had lost two front teeth, and I was not laughing anymore. Every time I smiled, I saw the giant hole between my teeth. Due to that setback, I could not enroll in time for the next semester. I contacted my dean, and he told me it was better returning to Puerto Rico. They had already made the necessary arrangements so that instead of returning to the General Studies Department, I would return to UPR but as Professor of Physics at the Faculty of Natural Sciences instead of my old job. I told Dr. Facundo Bueso that being part of the faculty was my greatest wish. Now I could teach physics, which is more difficult than teaching physical science as it requires a better mathematical background. For me, that was no problem because I was so experienced in all branches of mathematics. I honestly was blessed with that change that came from nowhere. So God was on my side.

PROFESSOR OF PHYSICS AT THE UPR

My departure from politics was just on the surface because, when I returned, my old political buddies came to seek my services. I agreed to stay as a consultant on political issues, but I told my fellow politicians that my vocation was teaching physics and mathematics. I had already convinced myself that to be a brilliant politician, I would have done things I never wanted to do. The worst thing that can happen to a person is to have political enemies. So gradually, I began my transition from politics into the academic environment. I had already studied the historical background of Puerto Rican politicians and began to lose interest in politics. I convinced myself that the solution to the political status of the Island was futile, at that time, and would remain so for many years to come. Today, the political status of the Island has not changed and probably will never change. So I was right in 1963.

As a faculty member, I was in charge of teaching Physics 101-102 and their respective laboratories. I also continued teaching physical sciences in the Extramuros division and mathematics at night. Again, I was receiving three checks. I had already recovered from dental problems, had false teeth and a new smile. This helped me a lot in the field of love. At that time, I had many girlfriends and was kept pretty busy with friends and academics.

One day something catastrophic happened. Dr. Facundo Bueso and Prof. Kiko Pesquera went out fishing. The boat was quite far from the shore. They say that Dr. Bueso was swimming under water, and when he rose to the surface, the boat propeller cracked his skull, and he died instantly. The University mourned for weeks. For me, it was an especially heavy blow because his sons were my friends. One of them was my classmate, and I had met the other son at parties in Dr. Bueso's house. Besides being my boss, Dr. Facundo Bueso was at one time my physics teacher.

Don Jaime Benitez changed the name of the College of Natural Sciences Building to the Facundo Bueso Building. At the same time, Don Jaime Benítez made a number of administrative changes. After a few months, everything returned to normal. What I liked about the administration of Don Jaime was the concept that the UPR was a house of studies with a strong internal autonomy. The strongest supporter of this concept was Don Luis Munoz Marin who was then the governor of Puerto Rico. He never interfered in the affairs of the university. Don Jaime was at the helm and kept the ship in the right direction. I have to reiterate that the people of Puerto Rico owe an extraordinarily large debt to Don Jaime because UPR under his management rose to the highest academic level. Improving the physical facilities and the teachers was the cornerstone of his administration. Of course, nothing in life is perfect. I understand that there were clearly some flaws during Don Jaime's administration that deserved more attention from him.

One day, the dean called me into his office and asked me if I wanted to return to Columbia University. Once again, I immediately answered with a big "Yes." In the summer of 1964, I returned to New York to continue my studies toward a doctorate in physics. This time I was better prepared and more mature emotionally. I already knew what to expect. Previously, I lived in the college dormitories, but this time I decided to rent a one-room studio apartment which was centrally located. I was lucky to find an apartment that only cost $75 a month and was near Riverside Drive and 96th street. From my apartment, I could see the Hudson River, New Jersey, and the George Washington Bridge. The George Washington Bridge is a monumental bridge that spans the Hudson River. It connects New York City with New Jersey. Sometimes I just sat there watching the traffic across that bridge. The bridge started in the Island of Manhattan where I lived and went to the other side around Fort Lee, New Jersey. It has an upper level with four lanes in each direction and a lower level with three lanes in each direction. There is a path on both sides of the river for pedestrians and people on bikes. At night, the bridge offered a spectacular view.

The subway was a block away. The apartment site was considered exclusive. The small apartment was freshly painted and had a stove, fridge, heating, air conditioning and a splendid view. Although it was small, it certainly met all my requirements. I cooked my own meals, and I was close to school. It was then that I decided to work with all my might in pursuit of the doctorate.

I had always been a big fan of both baseball and boxing. So on weekends when I needed to clear my mind; I went to watch baseball games at New York's Yankee Stadium and Ebbett's Field, the Brooklyn Dodgers Stadium. I enjoyed going to the stadiums because they reminded me of the famous players who played there—Babe Ruth, Ted Williams, Joe DiMaggio and others. Besides that, the New York Yankees baseball club has always been a formidable team, and the best players played and still play for that team. One day it occurred to me that the games shown on TV were free. All I needed was a decent antenna and a TV set. One day I started looking in the trash where people threw old electronics and found a 27-inch TV set. It weighed a ton. Since I knew electronics, I fixed the TV, and in no time I was watching TV in my apartment. The only problem was that the TV was on the chassis only. The wooden cabinet was missing. The bared TV was my only source of entertainment. I lived by myself so the missing TV cabinet did not bother me. On the contrary, since the TV was all vacuum tubes, it kept my apartment warm during the winter months. During the summer, I had to open the window to let fresh air enter and chill down the TV and my apartment. Since I lived close to the Hudson River, the apartment was cooled down by just opening the window. No air conditioning was truly necessary.

Some weekends I would go to watch boxing live at Madison Square Garden where one could watch the best boxers of the time. Boxing was also shown free on TV. In those days, there was no cable. All TV was free! Going to

Madison Square Garden was like entering a sacred site. The atmosphere and the events that one watched were worth the money paid.

At other times, I just walked around the downtown area near the famous 42nd Street and Broadway. In the bars and nightclubs, one could hear the best jazz bands. There was always a section where one could sit, drink a beer, and listen to Stan Kenton, Dizzy Gillespie, Billy May, Gene Kruppa, Louis Armstrong and other jazz greats. I paid for only one drink at the bar and then I sat back to listen for a couple of hours. Nobody cared. Sometimes I would see the magnificent Latin bands such as Arsenio Rodriguez, Tito Puente, Tito Rodriguez, Machito, Perez Prado, Xavier Cugat, and others. New York has always been the capital of music of all genres. The best recording studios were located there. You could see the magnificent offices of the most famous record labels such as; RCA, COLUMBIA, and DECCA. I was lucky to be there in the early 60's and able to enjoy the best musicians and orchestras.

My biggest surprise was that one day I was walking down 42nd street and saw a rather large sign in front of a theater that read: "ELVIS PRESLEY IN PERSON." I did not think twice. I immediately bought a ticket to see Elvis Presley live. I had already listened to Rock 'n Roll albums, but it was never the same as seeing it live, watching the performance of the musicians and singers and the screams of all the girls going crazy over Elvis.

Another day I almost lost my mind when I saw a sign that read: "THE BEETLES LIVE." I had already heard their music but seeing them in person was a must. This group of British boys broke all box office records and was the greatest sensation in the music of Rock 'n Roll in the whole world. For the first time, Elvis had competition. We woke up with Rock 'n Roll and went to bed with Rock 'n Roll. It was played on the radio, on TV, and even on public transportation. I was fortunate to be living in New York at the time that music made history around the whole world.

One day I went to visit my uncle Matias who was a cook at the prestigious Waldorf Astoria in New York City. While talking to my uncle in the kitchen, sounds coming from outside caught my attention. I asked my uncle what all the fuss was about. He told me that the Perez Prado Orchestra was rehearsing in a room that was adjacent to the kitchen because Xavier Cugat was using the facilities at the penthouse. I was frozen to the spot. I had seen the orchestra before in person in Puerto Rico in 1955 with Hispanic musicians performing in small theatres in Santurce. What a thrill it gave me to see and hear again live the Perez Prado Orchestra 10 years later in New York City! I even had the opportunity to watch a rehearsal free of charge and hear some of the best American musicians like Maynard Ferguson and Doc Severinson on trumpet. Summoning all my courage, I asked Perez Prado to tell me the secret of his orchestra's unique sound. He explained to me that the new sounds actually came to him by listening to the sounds in nature, like the song of a bird, or the sound of water falling on the surface of a rock. He also told me that he only allowed the best musicians to play in his orchestra. So it was a veritable United Nations consisting of "gringos," Mexicans, Cubans, Puerto Ricans, and others. I have never forgotten that meeting with the Maestro. At the time, I did not know that destiny would have us meet again in the future.

Despite all the events that I have just described, I was still totally focused on my graduate studies. The biggest problem I had at that time was the Vietnam War. At Columbia University, there were student protests every day around noon. Students were chanting in the center of the University close to the library and also burning their draft cards. I saw that FBI agents were taking pictures and movies. I did not want to involve myself in the protests because my mission was to study. As American citizens, Puerto Ricans are also obligated to serve in the military. At that time, military service was compulsory from age 18 to 35 years old. The Selective Service Board was responsible for enforcing that law. If someone broke the law, then the case went to the jurisdiction of the FBI. I was classified 1A so I would have been among the first to go to war. So instead of joining the student protest, I went instead to the Selective Service Board office and formally requested changing my status from 1A to 4F. I gave them my reasons for the request, and about three weeks later, I received my change of status. The Selective Service exempted me for a year so long as I continued as a Professor of physics at the University of Puerto Rico. I found that there was no need to stand in public burning your draft card, unless, of course, you wanted to protest publicly.

MY FIRST AND ONLY MARRIAGE

In the summer of 1965, I returned to Puerto Rico to teach a summer physics course at UPR. Prior to this trip, I had corresponded with a girl named Nora who lived in the jurisdiction of the town of Naranjito. My brother told me about Nora, but I had only seen her in a photograph. I sent her my picture also, so she and I had an idea of what we looked like.

After I landed in Puerto Rico, I reported to my supervisor at UPR. My dean had assigned me a course in general physics with a laboratory. I had taught that course many times, so no preparation was needed. Then I sent a note to Nora that I was going to meet her next Saturday. I already knew that her father had been divorced from her mother. He lived in the countryside near the small town of Naranjito with his new wife and two daughters. Nora's mother, who had also remarried, lived in Rio Grande on the northeastern side of the Island.

The Saturday that I was supposed to go there, my car broke down. I did not have any way of getting in touch with her. Of course, there were no cell phones back then, and, unfortunately, there were not any public pay phones around either. I was extremely upset that she and her family would think badly of me. I asked my friend Gilberto Cruz to accompany me on my next Saturday trip. Gilberto and I traveled to Naranjito to meet Nora, her father, and the rest of the family. I made that trip from my hometown to Naranjito for several weeks.

I also had the opportunity to meet her mother who lived on the outskirts of the town of Rio Grande. I made several trips there to meet the rest of the family.

I had to return to New York sometime at the end of August, so I told Nora that in order to formalize the relationship; I would buy her an engagement ring. Her father was not happy with that. He told me that I had two choices. (1) Get married or (2) not return to see his daughter. I took the second option as an ultimatum. I spoke with my family to see what their opinion was. My brother-in-law, who was an immensely wealthy businessman, told me that he would take care of the wedding expenses. He gave the order to his accountant to pay all bills, including a stay in a Barranquitas hotel and a new hotel in Fajardo, a famous eastern port town. Hotel El Conquistador was then, and still is, a 5-star hotel. On Saturday, I told her father that the wedding had to be before September because I had to return to New York to pursue my graduate studies.

The wedding took place on August 28, 1965. This wedding became an immensely significant event for all the people involved. I rented a prominent place in order to accommodate so many people. I had to make sure that not only the relatives of the bride be included, but also my relatives who were a large group. To add to the excitement, that weekend a powerful hurricane was expected to hit the Island. I remember that at one point during the reception, the electricity went out for an hour due to a power failure. Despite that, everyone who attended the wedding party had a fantastic time. There was dancing, food, drinks and the threat of the hurricane, which was a freebie.

A week later, I was in New York with my bride. Because the marriage took place with the speed of lightning, I did not have time to say goodbye to my mom. I never knew if she even attended the wedding. She never expected me to marry that quickly. My mom kept a photo album of all my girlfriends, but Nora was never in that album. Maybe she expected that I would marry a woman of her choice, but I would never have agreed to that.

My wife's name is Nora Morales Mercado, and from this union came three boys. The first, Juan Carlos, was born on February 21, 1967; the second, Ricardo, was born on May 13, 1969; and the third one is called "the baby," but his real name is Eduardo Gabriel, born on July 7, 1987—20 years after Juan

Carlos. When Eduardo was born, I was 50 years old, and my biggest concern was wondering if I would live to see his graduation from at least the eighth grade. I have lived not only to see him graduate from eighth grade, but also from high school. A year later he earned his Baccalaureate degree from De Paul University in Chicago, Illinois. Then a year after that, I saw him receive his Master's Degree from Northwestern University in Chicago, Illinois. I was also very blessed to be able to attend the commencement ceremonies of my other two boys. Ricardo earned his Baccalaureate degree from the University of Central Florida, and Juan Carlos earned his Baccalaureate degree from Florida International University in Miami, Florida.

My three children have become professionals, and they probably make more money than I earned when I was working. I made many sacrifices so that they could become exemplary citizens and be able to earn professional degrees from recognized and accredited universities. Thank God that the three obtained their degrees and that God gave me excellent health so that I was able to attend all those graduation ceremonies. Marcel Rivera is my only grandchild. Marcel's mother had a relationship with my oldest son, Juan Carlos, while he was stationed in Germany. My oldest two sons served in the Air Force, and they are veterans of the Air Force.

THE VIETNAM WAR AND ITS EFFECT ON STUDENTS

The Vietnam War (the Pentagon called it a "conflict") affected the trajectory of my life. For the benefit of those who are not familiar with that war, and to understand the situation of many young people like me who were affected by this conflict, it is helpful to give a brief summary of what took place in Vietnam.

Vietnam was a French colony until the middle of 1950. Circa 1954, France decided to give Vietnam its independence. A treaty was signed in Geneva in 1954 giving independence to Vietnam, Laos, and Cambodia. Vietnam was divided into two republics. One of them was known as North Vietnam and the other South Vietnam. France's plan was to unify North and South Vietnam through elections to be held in 1958. Saigon was the capital of South Vietnam; Hanoi was the capital of North Vietnam.

The unification of the country was not achieved due to a coup led by Ngo Dinh Diem. He refused to hold the elections, denying any possibility of unification. Then in North Vietnam, Ho Chi Minh assembled a guerrilla group to attack and overthrow Ngo Dinh Diem. There was a lot of corruption in the government of Ngo Dinh Diem, and this started the creation of several guerrilla groups with socialist leanings. The most famous was the National Liberation Front—better known as the Vietcong—which was under the command of Ho Chi Minh. Due to the inability of the corrupt leader of South Vietnam, the U.S. President Eisenhower, decided to use the support of the CIA (Central Intelligence Agency) to overthrow Ngo Dinh Diem and install a leader which could manipulate the rebels. The logical consequence was the further division of the country. Ngo Dinh Diem died in 1963 due to a coup—allegedly sponsored by the U.S. government under the leadership of the CIA. Americans said that Van Thieu was a puppet. For the first time, Americans were faced with an enemy that had no uniforms or conventional weapons. Guerrilla warfare was not particularly well known by the American leaders. The Americans found themselves confronting an enemy that worked on the bases during the day, and then attacked the troops with guerrilla command strategies at night. Neither increasing the number of soldiers nor increasing the number of airplanes or helicopters proved sufficient to win the war against the Vietcong.

South Vietnam was in chaos due to the lack of experience of its inefficient leader. The U.S. President sent a group of Navy Commandos to help strengthen their soldiers because the Vietcong threat grew stronger every day. The Communist governments of China (under the leadership of Mao Tse Tung) and Russia (under the leadership of Josef Stalin) sent reinforcements to help the Vietcong. President Eisenhower sent more troops to try to stop the advance of the Vietcong. Since the beginning of 1960, the war continued to escalate. Years later, President Kennedy sent more troops and so did China and Russia. By the middle of 1964, thousands of soldiers on both sides had been killed, and the Vietcong continued its advance southward. Years later, President Lyndon B. Johnson escalated the war—ironically renamed the "Vietnam Conflict." The United States was losing the war and the prestige of being a world power. The American troops learned that the Vietcong was on the outskirts of Saigon when least expected. The United States, under the leadership of President Richard Nixon, had no choice but to destroy their own military equipment so as not to fall into the hands of the Communists. They fled Saigon and ultimately had to turn over the territory of South Vietnam to the Communists. Until that time, it was the first war that the U.S. lost. Ironically, 20 years later, another U.S. president repeated the same story in another part of the world. This time, the battleground was moved to Iraq, Afghanistan, and surrounding countries. Much the same happened between North Korea and South Korea, but it was the Vietnam War that affected me the most.

In 1965, the United States had a series of conflicts both on the domestic front and abroad. Students were protesting. The black community was rioting. More troops were needed to send to the front lines of Vietnam. One day I got a letter from the Selective Service Office to report immediately to X site to begin the process to enlist in the U.S. army. I had no choice but to show up before the Selective Service Board to appeal the decision. At the same time, I called my dean in Puerto Rico and told him I had to stop my doctoral studies because I had to go to Vietnam. The dean told me that teachers were exempted. All I needed to do was present a letter from UPR stating that I needed to return to the Island to continue my teaching duties. After submitting all the necessary documents, the Selective Service Board exempted me from military service on the condition that I report every year to the local Selective Service Board to see if the exemption had been extended for another year. If it had not, I would have to join the army. In May 1966, I had already completed all the requirements for the Master of Science degree and, therefore, could return to the Island to continue teaching at UPR.

HOW I EARNED SOME EXTRA MONEY

In August 1966, I returned to my teaching position at UPR as a professor of physics at the Faculty of Facundo Bueso. I shared an office with my colleague Prof. Felix Pereira who was older than I. By this time, my academic credentials were quite solid and he enjoyed sharing the office with me. Of course, as he was twice my age, there was no doubt that he had more experience than I had dealing with students. Therefore, Prof. Pereira became my mentor in student affairs. I learned to deal with all kinds of student problems. I became an expert lecturer and developed such an affinity with students so that wherever I was, the students sought me out. I never refused to answer any questions in physics and mathematics. I immersed myself in mentoring issues. Students wanted to pay me for my services, but I did not accept money or any reward that might stain my academic morals. For example, one day, a girl was with me in my office and asked what grade she had in class. I told her that she was failing the class, and I would recommend that she make a tremendous effort to improve her grade. Then she stunned me by asking what she could do for me so that she could get and an "A" in the class. Sure, she was quite lovely and had a beautiful body. She looked like a Hollywood artist. On top of that, we were alone, and the only thing I had to do was close the door to my office. Thousands of things went through my mind. She kept looking at me, waiting for a response. Then I said: "Miss, I'm a married man. My wife is prettier than you and gives me everything a man could need. I cannot corrupt myself and violate academic rules and those that society imposes on me." She looked down then left my office. I went home and got drunk.

I must confess that similar situations happen at all schools and universities in every country. In fact, at that time, I knew several professors who had sex with their students. My advice to any teacher is to avoid being corrupted. We came into this world via a woman, and I have always believed that if you want to be respected by others, you have to first respect yourself.

When I lived in New York, somebody broke into my apartment and stole money and electronics. As the apartment was not mine, I designed a burglar alarm to protect the apartment entry door and windows. After the installation of the alarm, my friends all liked my idea and asked me to install alarms in their apartments. It was then that I got the idea to patent my burglar alarm. Instead of going to Washington DC, I did all the required paperwork at the New York City Public Library. The patent office had several district offices around the country. In the state of New York, one of the patent district offices just happened to be at the New York Public Library. I went there, and they helped me get the patent for my burglar alarm without having to deal with a Patent Attorney. I was protecting my idea thanks to copyright law.

Months later, someone stole my neighbor's car radio. He asked me if I could put an alarm on his car. Several weeks later, I designed a burglar alarm to protect any car. This idea was also patented. I installed two or three alarms for neighbors and friends but never thought about the commercial potential.

When I returned to Puerto Rico, some friends started asking me about alarm installation in cars and houses. Thieves have always been busy in Puerto Rico stealing from people's houses, businesses, and cars. It was then that I decided to open a burglar alarm company. As I was a UPR professor and did not have the time to walk the streets selling burglar alarms, my brother Pepito and my wife helped run the business. After the business was growing, I was forced to recruit more employees. I trained my employees as well as the alarm installers. I taught them everything they needed to know—except how the alarm worked. That was a secret, and I was the only one who installed the alarm brain.

I needed to involve myself more in the business, but that jeopardized my status as a university professor. My company was called ALERT ALARMS. My competitor was ADT (American District Telegraph) which was, and still is, one of the largest companies in the field of security. My alarms were installed in jewelry stores, banks, pharmacies, rich people's houses and all kinds of retail stores. My company covered the entire island of Puerto Rico, and it was a lot of work for me. The business grew so much that it could not cope with the demand. Apparently, the thieves were just as busy as I was!

I bought several houses, a three-acre farm, several cars and business trucks. I had savings and checking accounts both for business and personal. All that wealth was too much for me. Everything in life has a positive side and a negative side. In my case, the downside was not having time to spend with my wife and my family. With money also come vices. I never smoked, but I was becoming an alcoholic. I never wanted to get involved in any kind of drugs. In addition to my alarm business, I learned how to make wine on a large scale. I also learned how to make rum although I never got into the alcoholic beverage business. From 1966 to 1969, I made more money in the burglar alarm business than I did by teaching. On the positive side, the burglar alarm business put me in touch with the rich and influential people of Puerto Rico. My clients swore that I was Cuban. Most of my clients were Cubans who had arrived in Puerto Rico from 1960 to 1970. They once had businesses in Cuba. In fact, the owner of the CMQ, its staff, artists, and musicians had settled in Puerto Rico. Mr. Mestre, who owned the CMQ in Cuba, became my friend. His son was one of my students at UPR. The WAPA TV station was controlled by Cubans. At that TV station, I met many artists from soap operas and music. The list is long, but I knew the members of Los Guaracheros de Oriente, La tremenda Corte, El show de Olga y Tony, Benny More, Perez Prado, La Sonora Matancera, Gaby Fofo and Miliki, Armando Roblan and many more. I met Cuban traders, doctors, jewelers and so on.

One day, Tito Rodriguez (who was Puerto Rican) called me and asked if I could install a burglar alarm at the Pagoda house he was building in Santurce. Yes, the house was shaped like a pagoda because Toby, Tito's wife, was Japanese. I installed the security system at that house with my own hands. It was the same security system that I installed in banks. It was a success. Hence the word spread like wildfire among the show business people, and I had to install alarms on the properties of other well-known artists such as Johnny Rodriguez who owned "El Cotorrito." Then I installed the burglar alarm at the Association of Artists of Puerto Rico which had a property in Santurce, run by Lito Pena and Sammy Velez. Some other people I met there were Damiron y el Negrito Chapusaux, Sylvia de Grasse, Cesar Concepcion, Rafael Cortijo, Ismael Rivera, Mario Hernández, Felipe Rodriguez and many more. I enjoyed going to that property because from the moment I entered until I left, I never stopped laughing, listening to all the stories and antics of these artists. People know the artists on the stage, but when one is with them in person, they are different people. I was lucky to have had the opportunity to be around those show business people.

Tito Rodriguez asked me to be his audio engineer for local shows. He needed a reliable audio engineer for live shows, especially those around the Island where the stage and all the electronics had to be set up. That was not an easy job. I told him that I would do it only in the San Juan area but not inside the Island. I honestly was doing it as a favor because I already had too much on my plate. One day, Tito asked me if I could be part of his staff in a show that he was doing in New York. I told him that I could not be absent from Puerto Rico even for a day. It was then that I began to withdraw from the artistic scene because, besides the long nights, there was the drug problem, and I had never done drugs. I could not stay late at night because I was a university professor and I had to keep myself sober. To complicate matters, some people were homosexuals, lesbians, drug addicts, and alcoholics. I didn't enjoy this environment and decided it was best to distance myself from that kind of work and not fall victim to the vices. It was hard for me to make that decision because I cherished the friendship of many of those people.

STUDENT PROTESTS PARALYZED UPR

In 1968, the University of Puerto Rico was a political hotbed due to the politicization of the two groups for and against the ROTC on campus. At that time, the UPR chancellor was Abraham Diaz Gonzales and, under his administration, student protests increased. Between the years, 1967 to 1969, I remember a series of confrontations between ROTC cadets, and AUPE and FUPI. In Rio Piedras, several students were injured, and the ROTC cadets' parade was suspended. There were a number of fights between the annexation group (AUPE) the pro-independence group (FUPI) the police, the ROTC, and classes were suspended for a few weeks. A female student and a taxi driver were killed. Several people were shot and wounded, and a bomb was found in the tower of the UPR. Days later another bomb was found in the ROTC building. I had students who were on both sides of the issue, and they advised me not to give classes in the week that the protests would occur. They said that the police were armed to the teeth, and they had reliable information that they were planning to invade the UPR buildings searching for pro-independence and FUPISTAS. I went to my dean, and he said that until the UPR chancellor gave the last word, it was my duty to teach. As it turned out, I did not have to teach because the classroom doors were locked. All these incidents convinced me that it was best to leave the country until this issue was resolved. During the same period, there were other universities in and out of Puerto Rico experiencing the same problems. The Vietnam War had kindled discord against the United States and its president. Student protests were the order of the day.

While all this was happening, I was still a professor at UPR and was also involved in all matters that required my attention to my burglar alarm business. None of my fellow teachers knew I was living a double life. I complied with everything that was related to teaching activities, but I was also running my burglar alarm business. One day, Prof. Juan Luis Brusi, who was a lecturer at the Faculty of Humanities, invited me to join him on a school cultural trip to Mexico. He explained that he was the leader of the group and UPR would pay all expenses for him and his faculty assistant. He told me that all I had to do was to serve as one of the group's chaperones. The group consisted of 30 students, and a teacher was needed for every 10 students. He was the head of the whole group and needed another teacher who was young and could survive the trip. Since I had all the qualifications, after consulting with my wife, I told Prof. Brusi that he could count on me. As part of the requirements, I had to attend lectures one day per week for six weeks to study Mexican culture before embarking on the trip. I had to put my business activities on hold and study every detail of the history of Mexico; from the time that Hernan Cortez first set foot in that country until the 1910 Mexican Revolution. We used a book called *The True Story of the Conquest of Mexico*. This is a thick book, but I recommend that you read the first part of it.

On the scheduled day, the whole group flew from Puerto Rico to Mexico via Merida, which had an international airport in the Mexican Yucatan peninsula. When we arrived in Mexico, many students succumbed to Montezuma's revenge. Luckily, faculty members had no such problems because we drank bottled water or beer with food. We had to get used to new foods and a new environment. For me, it was like going back to the Puerto Rico I knew in my childhood. During the stay in Mexico City, only the best five-star hotels had bathrooms with hot and cold running water. Air-conditioning was a rarity. The hotel rooms had only fans to cool down the room. Outside the capital, hotels that we stayed in did not have bathrooms. Instead, latrines were placed behind the hotel. The showers were in a large room shared by all guests. I had to get used to bathing in front of other people. It was the same thing in the women's showers. There was no refrigeration. When the bus made a stop, the only refreshments available were at room temperature. But my father once told me, "When you go to Rome do as the Romans do." So I got used to the new environment. Prof. Brusi told me that that was one of the main objectives of this trip. We were studying a different culture, a different way of life. To be honest, I enjoyed the trip because in your own country, people can bitch about any issue, but when you travel to another country, you realize that you are living in heaven.

For three consecutive days, we studied the architecture of historic buildings and other points of cultural interest in Mexico City. We went to museums, cathedrals, churches, monasteries and the Pyramids of the Sun and Moon at Teotihuacan. We saw the extreme poverty in Mexico. The majority of Indians and the majority of Mexicans lived in what we considered poor conditions. It reminded me of when I was growing up in my old neighborhood. At that time, the Mexican population was divided into two classes of people—the rich and the poor. There was no middle class.

After three days in the Aztec capital, we left Mexico City unusually early in the morning and headed east. The next goal was to reach the Yucatan Peninsula. We spent one morning in the city of Campeche. In the afternoon, we reached the port of Vera Cruz. We again studied the points of interest in those two cities. Next day, we headed toward Chiapas, then Tusla Gutierrez to study the pyramids of the Maya Indians especially the Palenque, Uxmal and Chitchen-Itza pyramids. Then we went to Oaxaca for several days and saw many churches. In Chiapas, they have one church for every day of the year. That area has more than 300 beautiful churches, some with gold-plated altars.

Days later we went to Tehuantepec and saw many lakes in that area. We saw the ruins of the Toltec and Chichimec Indians. Anthropologists were still uncovering the ruins of the ancestors of that area. After studying the southern part of Mexico, we continued on the cultural journey.

Days later, we headed west en route to Acapulco. We went through many villages and met extremely fascinating people. For example; we stopped in Guadalajara one day. The brilliant and beautiful colors of clothing were extremely eye-catching. Next day we stopped at Taxco. We visited the mines where silver and onyx stone are extracted. We admired the art of carving in onyx and all kinds of jewelry made from silver. Everybody bought rings and other types of jewelry made of silver at rock bottom prices.

Finally, the next day we arrived at the beach in Acapulco. At that time, there were many hotels and other attractions for tourists. We saw where the Mexican rich people lived. This area was called the Mexican Riviera. We went to see "El Clavado"—a death-defying show. From a high rock, children looked down at the tourists who would throw coins into the water. Then the children jumped from that 100-foot-high cliff down to the water diving under to look for the coins. The child who got the coin kept it as a token for his show of manhood. It is a belief that if they can find the coin under water, then they will grow to be a strong man. It was remarkable watching these children risk their lives for a simple coin. Fortunately, people tipped them well to take their picture.

Speaking of photography, I had always been a photography buff, and I always had professional cameras. At that time, I took my most expensive cameras to document that trip, my first outside the United States. On that trip, I used an 8 mm (millimeters) film camera. Later on, I bought a Super 8 mm film movie camera. I saved those movie films for many years at the correct temperature so that they lasted longer. When those films, both 8 mm and Super 8 mm from all my trips, were getting old, I transferred them into the new VHS video format. Years later, the VHS video format became obsolete, and I again transferred all of them into the newest digital technology. Today, I still have all those videos that I took in 1968-69 in the new digital format. I tell my students that no matter how old I get, I always keep myself up to date in technology.

Finally, the day came when we had to return to Merida, Yucatan. Then we flew to Puerto Rico. I should point out that I have made many trips since then to Mexico. My most recent trip was in 2012. Merida Airport was the international airport at the time. Today it is a regular airport, and the Cancun Airport is now the international airport of that part of Mexico. Altogether, we spent about 20 days on my first Mexican trip in 1968. All of those students behaved exceptionally well. Felipe, the bus driver, became part of the group. Honestly speaking, our lives were in his hands. The area of Oaxaca is truly rural, and there are many dangerous roads. The large bus that Felipe was driving was extremely hard to keep on the road. The margin of error is probably a foot or two. Felipe was also the group bodyguard. We all learned a lot about the history and the culture of Mexico, so much so that most of the Mexicans I've met in my life have not known the country as well as I.

For those who do not know, I'll give a brief summary of Mexico. The official name of the country of Mexico is the United States of Mexico. It is a federal republic located in the southern part of North America. It consists of 31 states and a Federal District, where Mexico City is the Capital of the Republic. Mexico City happens to be also the capital of the state of Mexico. Yes, it looks confusing, so please look at the following table:

State	**Capital**	**State**	**Capital**
Aguascalientes	Aguascalientes	Morelos	Cuernavaca
Baja California	Mexicali	Nayarit	Tepic
Baja California Sur	La Paz	Nuevo León	Monterrey
Campeche	Campeche	Oaxaca	Oaxaca
Coahuila	Saltillo	Puebla	Puebla
Colima	Colima	Querétaro	Querétaro
Chiapas	Tuxtla Gutierrez	Quintana Roo	Chetumal
Chihuahua	Chihuahua	San Luis Potosí	San Luis Potosí
Distrito Federal	Ciudad de México	Sinaloa	Culiacán
Durango	Durango	Sonora	Hermosillo
Guanajuato	Guanajuato	Tabasco	Villahermosa
Guerrero	Chilpancingo	Tamaulipas	Ciudad Victoria
Hidalgo	Pachuca	Tlaxcala	Tlaxcala
Jalisco	Guadalajara	Veracruz	Jalapa
México	Toluca	Yucatán	Mérida
Michoacán	Morelia	Zacatecas	Zacatecas

Note that each state has its capital. The DF (the Federal District) is a state, and the state capital is Mexico City, which, in turn, is the capital of the whole republic. There is another state called Mexico and the capital city is Toluca. This causes much confusion among many people, especially poorly educated Mexicans. For me, this trip changed the path of my life. When I returned to Puerto Rico, my life took a new direction.

TOWARD MY DOCTORATE IN MEXICO

After my return from Mexico, I resumed my job at UPR. I then did a review of the balance sheet of my burglar alarm business. I also did an analysis of my personal life and my professional life. I concluded that I had to choose between the business and teaching. The decision was not easy for me because I made more money in my burglar alarm business than in my teaching job at UPR. I knew that businesses also have some inherent risks, and I had heard about other businessmen who had gone bankrupt because of lack of capital or incompetent management.

One morning Dr. Curet called me into his office and asked me if I wanted to finish my Ph.D. Without hesitation, I said yes. I told him that my preference was the National Autonomous University of Mexico (UNAM) in Mexico City. He told me it was okay because there was another professor from UPR working for his Ph.D. in physics at that university. I did not know that Prof. Mattos Cintron was in Mexico. Dr. Curet told me that I should use my sabbatical year to pursue my graduate studies. This needs an explanation. When a teacher is on sabbatical leave, the institution pays a regular salary to the teacher as if he were still working. The objective is to enable a teacher to spend all his time working on a thesis or a book. Another choice is to continue the teacher's education at a university abroad to strengthen his mastery in his own area of interest.

In May of 1969, I made the necessary arrangements to move to Mexico. I did not want to sell my belongings. I just wanted to move them from point A to point B. The difficulty was that point A was Rio Piedras, Puerto Rico, and point B was the capital of Mexico. I had to get all the documentation required for me and my family in order to travel to Mexico. Unfortunately, there was no Mexican embassy in Puerto Rico; the closest one was in Miami, Florida. The biggest problem was how I was going to transport my belongings from one place to another.

One day I was driving along Fernandez Juncos Avenue. I saw a man trying to load a grand piano onto an enclosed trailer that had a large door in the rear. It caught my attention, so I parked my truck in the store parking lot. I asked the owner, a Cuban businessman, where he was taking that piano. He told me that he had another piano store in Miami and was transporting the grand piano to the San Juan shipping port. I told him I was interested in buying the trailer because I was moving to Mexico and needed it to take my belongings to Miami. After some negotiation, I bought the trailer for $800. After further negotiation, he charged me $200 to bring the trailer to a barge. The barge destination was the Port of Miami. He also told me that when the trailer with its load arrived in Miami, he would help me get the proper port clearance because he had the necessary connections at the port. I did not have to pay any taxes as long as the shipment was destined to his warehouse in Miami. I paid him $1,000 cash.

A week later, all my household furniture was in the port of Miami, including the trailer and my truck. I left everything in storage in the Miami warehouse and then flew back to Puerto Rico to pick up my wife and my two children. I flew back to Miami to claim my property that was housed at the warehouse. It was a bargain moving not only my furniture, but also the trailer and my truck from Puerto Rico to Miami. At that time, I owned a brand new 1968 four by four GMC Jimmy. The entire move cost me only $1,000. If I had used the services of a moving company, it would have cost me over $3000. For safety reasons, I never told anyone the final destination of my possessions. I had already rented an apartment in Mexico City, which was located in an exclusive area called "El Pedregal." Only wealthy people lived in that area, but I was not aware of that. The owner of my apartment was a Spanish businessman who had settled in Mexico.

I left on a Monday early in the morning from Miami, Florida, en route to Mexico by road. First I had to cross several southern states—Miami to Tallahassee, then west toward the southern parts of Alabama, Mississippi, and Louisiana, finally crossing into Texas towards San Antonio. The plan was to enter Mexico at the border city of Laredo, Texas. I spent five days crossing the southern part of the United States until I reached Laredo. I had to do all the driving because my wife did not want to drive a vehicle with four-wheel drive, which was pulling a trailer that weighed several tons. Thank God that gasoline was cheap during that time. Driving from Miami, we made the first stop in Mobile, Alabama. It was there that I noticed that the spare tire of my trailer was missing. Apparently it had fallen off somewhere in transit. I decided not to waste time looking for a spare tire because they were uncommon tires which could only be purchased through a regional tire distributor, the nearest one being in San Antonio, Texas. Therefore, I had to risk continuing the journey without a spare tire for the trailer. My truck had a powerful engine, and I had made some modifications to both the vehicle and the trailer to avoid any problems on the road. I installed a cooling system for the transmission because transmission oil changes its viscosity at high temperatures. I put "anti-sway bars" on the trailer to prevent a trailer jack knife. Many drivers have lost control of a vehicle pulling a trailer and the consequences can be deadly. My advice to anyone who is pulling a trailer is to take all necessary precautions to avoid an accident. For example, if one brakes suddenly, the brakes of the trailer must be synchronized with the vehicle brakes to prevent an accident. It is beneficial to rest from time to time to prevent falling asleep at the wheel and causing a fatal accident. I drove for eight hours each day then stayed in a hotel to rest. Driving at 50 miles per hour (and when pulling a trailer, going faster than 50 miles per hour is asking for trouble) in eight hours one should cover about 400 miles.

When we arrived in Laredo, Texas, we had to cross into Nuevo Laredo, Mexico. At the border, the customs officer told me I could not enter Mexico because I needed an official seal (stamp) on my documents. Friends had warned me about all the things that could happen at the Mexican border, so the issue raised by the officer did not take me by surprise. I asked the customs guard how I could solve the problem. He told me that his boss's heart softened when he saw U.S. dollars. I gave him a $20.00 "gift" and he let me go. To enter Mexican territory, one had to pass three customs posts, and that was only the first one.

At the second post, the customs guards were already waiting for me. I approached the guard, and he told me, "Sir, you have to empty the trailer to see if you are bringing weapons." I replied, "Lord, I'm just a student who barely has enough money to live. I'm not a terrorist, and I have no weapons." The guard turned away from me and went to a booth, and about two hours later he came back and said, "I thought you had finished emptying the trailer." "Look sir," I replied, "My wife and my two children do not have the strength to help me. May I ask you how I can solve this problem?" He said, grinning, "Look, me and my buddies are paid a low salary. Any monetary help will be welcome." I gave them another gift in the amount of $40.00 and the police were quite happy. Then the guy looked at me, raised his voice and said, "I hope you have good luck at the third post." I made no comment and went on my way.

The next day I awoke early and approached the third and last post. Reading the face of the customs guard, I noticed that he was looking at his wrist watch. Then he said, "Look, compadrito, right now we are closed and you have to wait until tomorrow at 8 a.m. so that we can inspect your trailer." I froze. Then I saw a man dressed like a priest, and I went to him seeking help. "Holy Father, can you help me with my problem? The police will not help, and I cannot wait until tomorrow. I've always been devoted to San Antonio, and I hope that you and the saints in heaven can help me." Well, that man, who appeared to be a priest, looked at me with deep-set eyes, and, pulling at his mustache, said, "Cursed be the mother of all priests. I am not a priest; I am the head of this post." I started to cry and then the priest imposter asked me, "How much money do you have in your wallet?" I told him that I only had $60 left to cover the cost of my trip to the capital. Then the fake priest said, "If you want to get to the capital soon, then why don't you make a contribution to those poor cops who are standing there, and you will see that they will thank you and maybe let you pass." Well, I had to take the risk of giving them the $60.00 to see if they would let me go. The guard took the money and told me to leave as soon as possible. Later, I learned that these incidents at the border in Mexico are known as "the bite."

I want to make it clear that the alleged police corruption in government offices has always been a natural thing, not only in Mexico, but in many central and southern republics. They say that it is better to look the other way and get used to it. This evil is part of the culture; therefore, trying to change it would be futile. The money that I did pay "voluntarily" helped these people who, in my opinion, are in need. I seriously love and admire Mexico. Ordinary Mexicans are exceptionally talented people and are always ready to help visitors, no matter how foreign they look.

We continued the journey to the nearest town. At that time, the road was extremely desolate. There were no people or other cars. Occasionally, I saw a car pass and sometimes I saw people mounted on "burros." The journey from Nuevo Laredo to Monterrey State was without incident. Then I crossed the State of San Luis Potosi, the State of León, reaching Morelia and gradually getting closer to Mexico City. Pretty soon, I started to notice that the number of all types of vehicles was increasing. Mexico City is in a valley that once was a lake. The valley is surrounded by high mountains, and therefore, all the gases escaping from vehicles have nowhere to go. The amount of smog that is trapped in the air is phenomenal. To complicate the situation, Mexico City is at a high elevation. This means that under normal conditions, oxygen is scarce; therefore, it affects a lot of foreigners. I noticed that most people had rather wide chests, indicating that nature had changed the size of their lungs in order to survive at the high altitude. After several hours, we got used to the smog and the high elevation. As I was wearing contact lenses, my eyes became red. It was bad news because I was the leader of the pack.

My goal was to get to the warehouse where I had already made arrangements to leave the trailer. After that, I planned to look for a hotel until the day came to move the trailer to the parking lot of the leased apartment. With the help of strong Mexican arms, my stuff was carried to the apartment, and in one week, the apartment was ready. I had to take the trailer back to Querétaro and leave it there during the time that I would live in Mexico. Incidentally, the city of Querétaro was where the Mexicans beheaded Carlota and Maximilian, who was the emperor of Mexico at one time in its history. The town is about an hour from the capital and the cost of living was cheaper than in the federal district.

Living in a Central or South American republic is not the same as living in the United States or Puerto Rico. One has to get used to the language, the country, and its way of life. The Spanish spoken in Mexico has many variations, and Mexicans speak differently from Puerto Ricans, Cubans or Dominicans. The same words in one country have a different meaning in another country. So my wife had to be extremely careful when talking in the streets of Mexico. People with higher education also speak differently from the masses. We always had to be aware with whom we were speaking. The food was quite different from the United States or Puerto Rico. Tacos or enchiladas made in Mexico are different from the tacos or enchiladas made in the United States. To my mind, Mexican food in Mexico is tastier than Mexican food in Puerto Rico or the United States. However, in Mexico, one has to be careful when buying food sold in the streets or small businesses. Water can also be contaminated. A Mexican can drink a glass of water without serious consequences. The same glass of water drunk by a foreigner can cause diarrhea that can last up to three days. My wife always took medication that was an antidote for "Montezuma's Revenge." However, I never had problems with either water or food.

Rivera as a last name is particularly common in Mexico, so the Mexicans thought I was Mexican. My Mexican friends told me I was "Yucateco." People from the Mexican Caribbean coast of Yucatan look and talk like Puerto Ricans. Therefore, I passed as a Mexican—but only until I drove my truck. The license plates on my truck were from Puerto Rico, and the plates displayed the following phrase: "Puerto Rico Isla del Encanto." When I drove through the streets of Mexico, Puerto Ricans began yelling, "Arriba Puerto Rico!"

There were many Puerto Rican families in Mexico City who came there long ago. Mostly these families worked in radio, television or the music industry. Many businesses were owned by Puerto Ricans. You could go to Puerto Rican restaurants or dance halls and feel as if you were back in Puerto Rico. Many Mexicans loved the music of Puerto Rico and Cuba. They called this musical genre "tropical music." Daniel Santos, Bobby Capo, Rafael Hernandez, Pedro Flores, and Virginia Lopez were just some of the Puerto Rican artists working in radio and television at the time. The Mexican Trio Los Panchos always had a Puerto Rican lead singer. Therefore, Puerto Ricans were not considered foreigners for Mexicans. We were loved because the Puerto Rican artistic class had taken root over there. On the other hand, many Mexican artists had traveled and lived briefly in Puerto Rico. The list is long and includes Pedro Vargas, El Trio Los Panchos, Jorge Negrete, Pedro Infante, Javier Solis, Jose Mojica, Dr. Alfonso Ortiz Tirado, Juan Arvisu and many more. (Rafael Hernandez and Benny More were both married to Mexicans).

MY BEAUTIFUL AND CHERISHED MEXICO

The Universidad Nacional Autonoma de Mexico (UNAM) was remarkable because they accepted college students not only from Mexico, but also from Central and South America as well as the Caribbean area. At that time, the University had about 100,000 students. My first assignment was to connect with my colleague, Prof. Mattos Cintron, who would take me where the physics offices and the library were located. Together we toured the most important buildings of the University. The Physics Building had about eight floors and boasted the best modern classrooms and laboratories of the time. The College of Medicine was as large as a hospital. The Architecture Building was enormous. The football stadium was immense. In short, it was larger than either of the campuses of Columbia University or the University of Puerto Rico. In fact, everything at UNAM was either big or huge—except the tuition. The annual cost per student per semester was only $250 (in the early seventies). That was a gift when compared to the cost of tuition from Columbia University, which, at that time, was $5,000 per semester or $10,000 per year per student. (Fortunately, tuition had not been a problem for me as UPR paid all the expenses of tuition, books, room and board. I only paid for things that were not considered educational.)

Having come from Columbia University which was much more difficult, my first semester at UNAM was pretty comfortable. In fact, I became popular there with the Mexican professors because I helped them grade students' work for free. Also, I was one of the few students who owned a new four-wheel-drive truck and perhaps the only one in all of Mexico. Every time I parked my truck anywhere, people came to see it, including the police who were always asking me for the truck papers. They were curious about how I brought the truck to Mexico. They wondered if there was a bridge between Puerto Rico and Mexico! I would laugh and say "Si."

After two months at UNAM, one of my teachers asked me if I was interested in writing a thesis in nuclear energy. He explained that Petroleos Mexicanos (PEMEX) was installing a nuclear power plant in Querétaro, and they needed graduate students who could work there. The graduate students would instruct the Mexican staff on safety issues. (The former president of Mexico, Lazaro Cardenas, nationalized Mexico's oil in 1938. Thus, PEMEX was, and still is, owned by the Mexican government.)

As atomic physics was my major, I began working on a dissertation on nuclear energy. The thesis topic was related to the questions of "Safety in a Nuclear Power Plant." Ever since the development of the atomic bomb, there has been a lot of debate on the safety of nuclear power plants. Using a nuclear reaction as a basis for generating electricity was a big concern at the time. And it obviously still is today. The main problem is that a nuclear reaction has to be controlled precisely within certain limits, and the margin of error is very small. If the reaction gets out of control, it could be catastrophic. The media has always focused on accidents in nuclear power plants. I worked a few hours a week in an office that was provided to me in the Physics Building. They knew that I was a physics professor at the University of Puerto Rico. Without realizing it at first, I had connected myself with the Department of Physics at UNAM. This connection would facilitate my immediate progress in my doctoral studies. Writing my account while living in Mexico would not be complete without a discussion of the Mexican tourist attractions.

TOURIST ATTRACTIONS IN MEXICO

Living in Mexico for a long time was incredible. It was like being a long-term tourist! I would like to highlight some of the points of interest of Mexico at that time. They are probably the same ones that could be seen today—with perhaps a few changes.

The Pyramids of Teotihuacan are about an hour's drive from the capital city. It is one of the largest archaeological sites in the country, and, during its cultural zenith (150 A.D. - 450 A.D.), it was the world's largest city with a population of over 200,000 inhabitants. The site contains two large pyramids—the Pyramid of the Sun and the Pyramid of the Moon, plus many smaller pyramids flanking the Avenue of the Dead. The Temple of Quetzalcoatl is also worth seeing. Visitors can climb to the top of the two prominent pyramids. From there, one can see the whole site, the surrounding countryside, and mountains.

The Plaza de la Constitución (El Zócalo) is literally the heart of the city—and the country. This square is the largest public square in the Hispanic world. (Red Square is the largest and Beijing's Tiananmen Square is the third largest.) In the center of the square, there was a pole with a giant national flag of Mexico. As this was and still is in the middle of the city, there was always something going on there. I remember the native dances, cultural events, music concerts, and political protests. There was Mount Mercy (Monte Piedad), where one could buy or pawn valuables. In fact, it was in Monte Piedad where I bought a dining room set and some Mexican living room furniture for my apartment. The furniture was made from real Mexican wood. There was also "the Public Desks," where many businesses such as lawyers, doctors, printers, barbers, and others set up shop. There were no walls separating one desk from the other—just rows of desks lined up to provide services to anyone who came there. I used the services of the Public Desks whenever I had to fill out an official document for the Mexican government. Some of the films of Cantinflas were filmed there. The area is surrounded by several places of interest for visitors to Mexico such as The Metropolitan Cathedral, the National Palace and other churches. Also within walking distance are the Palacio de Bellas Artes, La Torre Latinoamericana and Alameda Park.

The Metropolitan Cathedral is one of the largest cathedrals in the city, and, because it was built over a period of about 250 years, it features different architectural styles. It is located on the north side of the Plaza de la Constitution. The Aztec Indians had once built a temple on that site. Hernan Cortez, the Spanish conquistador who conquered Mexico, sent his men to destroy the temple and build a Catholic church in its place. Many of the materials used to make the cathedral came from the destroyed Aztec temple. At the time of the Aztec Indians, Mexico City was a vast lake. Then came a drought and the lake dried up. Due to the nature of Mexico City's land, the cathedral is slowly sinking (like other historic structures such as the Basilica of Our Lady of Guadalupe). If one takes a walk in the vast interior of the cathedral, one will see many religious displays and examples of architecture. This cathedral owned one of the world's first vintage organs in the world.

The National Palace is the heart of the Mexican government. I had to go there for any official documents that I needed to deliver to the Mexican government. This magnificent palace contains many of the murals of Diego Rivera, one of the world's greatest artists. Although he was once considered a communist, he is revered currently, and his murals are national treasures of Mexico.

El Templo Mayor is part of the archaeological ruins of the Aztecs. The Temple, dedicated to the Aztec gods Tlaloc (god of rain) and Huitzilopochtli (god of war), is also a national treasure. This was the main temple and the focal point of the great center of the Aztec capital of Tenochtitlan which, at that time, was in the same place but in a chain of islands in Lake

Texcoco. Many of the materials of the temple were used by the Spanish to build the adjacent Metropolitan Cathedral.

The Palace of Fine Arts was completed in 1934 and stages some of the best performing arts in the city. The interior contains murals by Diego Rivera, Orozco, Siqueiros and Tamayo. Visitors to Mexico City can enjoy a performance of the Ballet Folklorico de Mexico at that location. Next to the Palace of Fine Arts is the Latin American Tower, completed in 1956 with 44 floors. When I lived in Mexico, this building was the largest building in Mexico and Latin America. Visitors could take the elevator to an observation deck on the upper level with 360-degree views of Mexico City, the Valley of Mexico, and the surrounding mountains and volcanoes (on clear days). Chapultepec Park was ordered built by Maximilian to connect his palace in Chapultepec with the national palace. Formerly, there had been a short road in the middle of the park so that the imperial coach could drive more comfortably. It was the first park in the city, dating from the sixteenth century. Near the Tower and the Palace of Fine Arts is Alameda Park, which was the first city park, and which also dates from the sixteenth century.

Chapultepec Park is the largest park (1,600 acres) in the city and the country. Many paths lead visitors through acres and acres of woods and lakes. The Chapultepec Castle is located on a hill in the park and was the home of Emperor Maximilian I and Empress Carlota, now open to visitors and home to the National History Museum. There are excellent views of the Paseo de la Reforma and the park. The building has several balconies and is reminiscent of the castles along the Rhine River in Germany. Also located in the park is a zoo, a lake with boat rentals, an amusement park in Los Pinos (the official residence and office of the President of Mexico), the National Theater, and several other museums, including the Museum of Modern Art, the Natural History Museum and Children's Museum. As UNAM has the best school of architecture in Latin America, it is not surprising that the buildings that make up the museum have won architectural awards.

The National Museum of Anthropology contains the largest selection of pre-Columbian artifacts and art of the various pre-Columbian cultures. The museum is enormous and is divided into separate rooms according to each of the pre-Columbian cultures. The central courtyard is famous for its large square concrete umbrella. The showrooms include those dedicated to the Olmeca, Chichimeca, Maya, and Aztec civilizations.

Xochimilco Lagoon is located in the southern part of the city; these are the last of the chain of lakes that were in existence at the time of the Aztecs. The Valley of Mexico was once a large lake with several small islands. Touring Xochimilco gives the visitor an idea of what Hernan Cortez saw when he arrived in Mexico for the first time. The area now contains a number of channels where visitors can ride in brightly colored gondolas. The channels are all that remain of the chinampas (floating gardens), the type of agriculture practiced by the Aztecs. I sometimes took my wife and children there to enjoy a day with mariachi music.

The Basilica of Our Lady of Guadalupe is another highly popular place in the history of Mexico. The Virgin of Guadalupe is the patron saint of Mexico. The Virgin Mary appeared to the child Juan Diego on the hill behind the church, known today as the hill of Tepeyac. The official religion of the conquerors of Mexico was Catholic. Today, the basilica is not only an impressive national symbol but also the meeting place of all Mexican pilgrims who come to the capital to pray, seek God, and ask the Virgin of Guadalupe to help them with their problems. Upon entering the basilica, one sees thousands of crutches and other artifacts that are left by the faithful as a sign of the miracles in which the Virgin had interceded. On the hill is also another slightly smaller basilica that was first built in the sixteenth century. The largest basilica was built many years later. Juan Diego's Apron, with the image of the Virgin of Guadalupe, can be seen in the large basilica. It is perhaps the most revered church building in Mexico and maybe even in all of Latin America. The site is particularly notable in Catholicism as it is the second largest religious center after the Vatican City in Rome, Italy. When I lived in Mexico, the largest basilica was sinking as a result of frequent earthquakes and poor ground.

Garibaldi Square is not far from the center of Mexico City. This square, known world-wide for its music, is where the mariachis gather to present their music to anyone willing to listen—and pay. I usually went there with my wife to hear the mariachis sing. There were also excellent restaurants serving exceptionally delicious food. At night was when the Plaza Garibaldi got even better. I would get a mariachi band to serenade my wife. Serenades are extremely important for Mexicans because it is the best way to express love to the person you love. There were several bars around the square. Some of these bars had fantastic live shows. Families from other parts of Mexico would come here to meet, eat, sing, and have a fantastic time. I must admit that I loved Garibaldi Square, and I still miss it because it was there that one truly "lived" the "ranchera" songs in the style of Miguel Aceves Mejias and José Alfredo Jiménez whom I often saw at the live shows that were presented in the bars. Some years later, I also saw the famous Javier Solis singing there. I feel sad that when I lived in Mexico, Jorge Negrete was dead. I could never see Pedro Infante because "Pedrito" spent his time making films and traveling abroad where he earned more money than in Garibaldi Square. Miguel and Jose Alfredo could almost get inside the music. When one of them burst out in song, it made your hair stand on end. It was an experience never forgotten.

La Zona Rosa is located near the historic city center in the northern part of the famous Paseo de la Reforma. Although this site had not been copied from any in Europe, strolling in this area brought to mind the Champs-Elysees in Paris. The Paseo de la Reforma Avenue is exceptionally attractive and, best of all, has no traffic lights. At the point where several streets meet, there is a large circle called a "Glorietas" or roundabout. In the center of every roundabout is a fountain or statue with some important rider on horseback. The area is bordered by the Avenida Chapultepec and on the other side Insurgentes Avenue and Florence Avenue. This area is quite busy both day and night. At night, the area comes alive.

In Mexico, prostitution is prohibited, except in the pink area, and every major city has a pink zone. Prostitutes have to meet certain regulations imposed by the Mexican government in order to limit the spread of venereal diseases and drug abuse. The first thing that struck me at the time was the prevalence of doctors in that area who treated venereal diseases. The next thing was that in Mexico, medicine was socialized. The Mexicans did not have to pay to go to a hospital or see a doctor. One could walk into any pharmacy and buy any medicine without prescriptions at rock bottom prices. I do not understand why medicine in the United States has become such a lucrative business, and why the government cannot implement socialized medicine as in Mexico, Cuba, and many other countries in the world.

DAMASO PEREZ PRADO AND JUAN GARCIA ESQUIVEL

At that time in Mexico, most recordings were based on the monaural concept. Stereo recordings had not yet become popular. I had studied this concept when I was a student at Columbia. Many students were building electronic devices for sound reproduction with the purpose of reproducing music in its entire splendor. RCA Victor Recordings Studios was trying to improve the recording and sound reproduction of live recordings.

One day I saw an ad that caught my attention. There was a school of engineers meeting to discuss the issue of stereophonic sound reproduction. The meeting was being led by Juan Garcia Esquivel who was both a pianist and an engineering student. In the conference room, there were around 100 people. The conference was quite entertaining, but what caught my attention was that one of the attendees was Damaso Perez Prado. At the end of the lecture, I approached Perez Prado and Garcia Esquivel. They were talking about how to record the sound of an orchestra to its fullest dimension so that when one played the recording, it sounded live. I told them that the secret was to have a microphone for each instrument. In Mexico, the recording studios used a single microphone for the entire orchestra, but that was not an effective method. Since Garcia Esquivel was an engineering student and I had knowledge of electronics, we got some paper and started to draw the electronic system requirements. We made a prototype which he took to the RCA Recording Studios in Mexico City. In one afternoon, we developed what is known today as a "mixer" or audio blender, an electronic device that combines the sound recorded by ten or twelve microphones. We were fools because we did not apply for a patent for this device. Garcia Esquivel regretted it more than I because his livelihood depended on this equipment. He was a pianist and an orchestra leader who later on became known worldwide. (A search on the Internet will confirm the musical genius of these two men.) What did I get out of it? I got the friendship of two musical giants, Juan Garcia Esquivel, pianist/arranger and Damaso Perez Prado also pianist and arranger—two musical geniuses, two pianists, two artists whom I've always admired throughout my life. One is Cuban-born, and the other is Mexican. Years later, Perez Prado became a Mexican citizen. His son, Perez Prado Salinas, inherited the orchestra after his father died.

The population of Mexico is divided into two classes: the poor (the majority) and the rich (the minority). At that time, there was no middle class. The wealthy class usually went to doctors outside of Mexico—a favorite place being Houston, Texas. The rest of the people did not pay for hospitals or health benefits. I remember seeing medical doctors driving old cars. In the U.S. and Puerto Rico, medical doctors drive luxury cars. Of course, there were some medical doctors who had their own private practices, but these were in the minority. Foreigners living in Mexico had to pay for medical services. For example, when my wife needed the services of a dermatologist, I had to pay. When I needed a dentist, I had to pay for those services.

THE 1968 TLATELOLCO SLAUGHTER

My colleague, Prof. Mattos Cintron, told me that at UNAM, there were a lot of students who were planning a rally to honor those who died or were wounded in the "Slaughter of Tlateloco." The irony was that I had run away from the student protests in Puerto Rico and New York, only to set foot in Mexico and find myself embroiled again in student protests. My mom used to say "when one does not like broth then they served you two cups." For those not familiar with what took place in Tlatelolco, let me relate what happened there.

The slaughter of perhaps thousands of spectators and students occurred on October 2, 1968. I arrived in Mexico a year later, and student discontent over that tragic event could still be felt. Many students organized a protest march against the Mexican government. There were some issues involved that were triggering dangerous reactions among the students. Those same concerns were prevalent at other universities throughout Mexico. The meeting place was the Plaza of the Three Cultures in Tlatelolco, Mexico City. The march had been preceded by months of political unrest in the Mexican capital. Not only at issue were the many problems of the forgotten Mexican poor class, but also the student repudiation of the Vietnam War. It was said that the communist party of Fidel Castro was behind all this foment. On August 27, 1968, there was a demonstration that attracted more than 200,000 students who merged in the Zocalo Square. This site, in the center of the city, is frequented by tourists and visitors from other parts of Mexico. The governor of Mexico was at that time Luis Echeverria Alvarez. They say he gave the order for the military to suppress the student protest. The police and the military used pressurized water jets and tear gas.

UNAM and other universities within and outside the republic are "Autonomous." The concept of autonomy means that the government cannot interfere in the administrative affairs of an autonomous university. This is to prevent the government from bringing in people loyal to them to run a university. In 1968, the Olympics were held in Mexico and President Gustavo Diaz Ordaz was determined to rid the streets of the capital of protesters. He did not want student protests to destroy the planned games. He also did not want to give these student protesters a platform to convey their message to the whole world. As a preventive measure, Díaz Ordaz ordered the military to invade the UNAM campus. The army invaded the school grounds and indiscriminately struck many students. In protest, the UNAM chancellor, Javier Barros Sierra, resigned on September 23, 1968.

The government ignored the chancellor's resignation, and the student protests increased in size. On October 2, 1968, after nine weeks of student strikes, 15,000 students from various universities marched through the streets of the city, carrying red carnations to protest the invasion of the UNAM campus. By nightfall, about 5,000 students and workers, many of them with their wives and children, gathered in the Plaza of the Three Cultures in Tlatelolco. There were armored cars and tanks fully armed and ready to kill if necessary. It was said that Mexican Governor, Luis Echeverria Alvarez, gave the order to kill. In minutes, the whole square was covered in blood, including wounded and dead women and children who had accompanied their loved ones. The slaughter continued throughout the night with soldiers raiding the apartment buildings adjacent to the square. Eyewitnesses said that the bodies were removed in garbage trucks and then buried in unknown sites in the middle of the night.

The next day, there was a brief official government explanation stating that the students were armed and shot first at the army, and the army and the police only responded to defend themselves. Luckily, I was not there, but I knew many students who had been injured in the slaughter, and they swore to me that no student was armed. I'm inclined to believe that the students did not have the amount of arms or the training that the soldiers had. There was no reason to kill so many people of their own race.

The role of the Mexican army is solely to secure the borders against a foreign invader. The president of Mexico is not a member of the army. In Mexico, the army is not linked to the government as in other countries. Another function of the military is to help people in times of earthquakes or hurricanes. It is also used to prosecute and try to suppress drug trafficking. The bad blood between the students and the military continued after the slaughter of Tlatelolco. When I went to school in Mexico, there was a time when students took over the rectory building and stayed for several weeks there. The classes were not suspended. It is necessary to point out that in the rectory building were housed all student academic records. Weeks later, the president of Mexico, Diaz Ordaz, gave the order to close the university. The army reoccupied the rectory building during the closure. Luckily, I had already completed my doctoral requirements, but the closure of the university would affect me adversely later on.

A few weeks after the university reopened, college students claimed that the soldiers had stolen many students' academic records. The government neither denied nor acknowledged that this had happened. Without a doubt, some documents were lost. Among those documents were my academic records as well as other students' records.

I wrote to my dean in Puerto Rico to tell him that the University of New Mexico in Albuquerque, New Mexico, had offered me a scholarship to do post-graduate studies in Astronomy. The dean advised me to go to New Mexico if I felt that my life was in danger in Mexico. Plus, getting a post-graduate degree would do no harm. He told me that the situation in Puerto Rico was still "hot." The consensus by the administration was that the University of Puerto Rico was also a victim of leftist political students. I told him that the same situation was happening at the University of Puebla in Puebla, Mexico, and at the University of Caracas in Caracas, Venezuela. It seemed that universities had become centers of protest against the "status quo," and the leaders were always left-wing political students or communists. The news spread in Puerto Rico that the UPR might also be closed due to student protests. It was said that Fidel Castro's government was funding socialist students.

In the 1970 elections in Mexico, a new president was elected. In Mexico, every president has a period of six years and cannot be re-elected. The grim news was that Luis Echeverria Alvarez was elected the new head of Mexico. The first thing he told the news media was that the government would seek out left-wing students and communists that were hiding in Mexican universities or on Mexican soil, and he would recommend putting them behind bars. I sensed that problems would remain at UNAM for a long time.

One day, I went to the rectory building for copies of my academic records. Bad luck! They could not find them. I went every day to see if my records had appeared but always received the same answer. I had already completed my doctoral requirements. The only thing missing was the degree certification. One day I told my wife that she needed to start preparing for another long journey. I had decided to move to New Mexico. After several weeks of preparations, one day, very early in the morning, I left Mexico City and headed for New Mexico. To be honest, I feared that one day a stray bullet could either hurt me or kill me. There is a saying in Mexico that "life is worth nothing." I did not share that philosophy. To me, life was, and still is, more valuable than anything else. My wife and my children depended on me. Without me, they would be lost.

POST GRADUATE STUDIES IN NEW MEXICO

The journey from Mexico City to New Mexico was not easy. First, I had to pack my belongings and get my legal documents in order to comply with Mexican law for a non-resident. The Mexican government zealously enforces immigration laws. At that time, anyone who left the country could not be trafficking in drugs, weapons, or other types of contraband; therefore, when packing, all boxes or equipment had to be properly labeled. Days later, I had to get my trailer that was stored in the city of Querétaro. I also had to bring the men who would be in charge of loading my belongings into the trailer. After completing the manifest of what was in the trailer, I closed it and locked it with a locksmith lock. My biggest problem with the trailer was the same as before. I still had no spare tire!

One day early in the morning, my wife, my two children and I left Mexico. My truck was pulling the trailer that was heavier now than ever before. Traffic in the capital of Mexico is extremely chaotic, but I was used to dealing with any problems that might pop up along the way. The first mishap took place next day in a town that did not even appear on my map. The trailer blew a tire. What I admire most about Mexicans is their desire to help people. Not only that, but they have the ingenuity to create something to solve a problem. A gentleman came to me and told me that I did not have to worry because he had the answer to the problem. It would cost only $10. He took the spare tire and put on what they called "un bigote." It is a method of vulcanization of the tire material with another piece of old tire. It worked perfectly; he had solved the problem. (I should point out that vulcanization is an ancient technique that was used also in Puerto Rico to repair a tire.) He mentioned that he had a tire from a farm vehicle that could be adapted to serve as a spare tire. That would also cost me only $10, but we had to wait four hours.

So I decided to stay in a small hotel, but I was told that the village had no hotels. The man said I could stay at his house in a room that he had for his children. He said he would not charge me for the room. He could not offer breakfast; nor were there any bathrooms for bathing. To avoid any misunderstand, I did not ask where they took their baths. Then I remember what my father told me in my childhood: "When in Rome, do as the Romans do."

The next day we were back on the way. My goal was to go to El Paso, Texas. To get there, I had to use another way out of Mexico. My objective was to reach "Ciudad Juarez" that connects to El Paso, Texas. That meant heading toward "San Luis Potosi" to reach the area of "Torreon." From there, I went down to "Chihuahua." We spent a whole day In Chihuahua because I wanted to see the widow of Pancho Villa, although I could not confirm if the woman was really his widow. I was very interested in visiting the house. The bullet-riddled car where Pancho Villa was massacred was parked right outside her home. It was a living memorial to her well-known husband. I took a detour to visit the poor abandoned widow whose suffering showed so plainly on her face. I knew that tourists who visited her house always gave her money. Therefore, I gave away $60, which, at that time, was a lot of money. We also took pictures of her, the car riddled with bullets, and the blood stains which could still be seen in the car. I also bought handicrafts and other souvenirs of that area of northern Mexico. After this visit, we continued on our long journey.

I still remember the Sierra Madre mountain range which runs from north to south, from Nogales to Guadalajara. These mountains are stunning. The mountains of Puerto Rico are tiny in comparison. Also on that route, I could appreciate the wilds of Mexico. We saw a lot of prickly pears and the maguey plant from which is made the famous drink, "pulque." I stopped to have a glass of "pulque." That drink is highly intoxicating, but ever since I was a boy, I had been able to tolerate the effects of alcohol without getting drunk. We drove for hours without seeing anyone on the road. That scared me because I was with my wife and two kids, and the pulque was starting to affect me. I had to stop when the engine of my truck overheated, and I looked everywhere but saw no one. Then, from nowhere, people suddenly appeared to sell their crafts. A boy tried to sell me an eagle whose wings were at least five feet from tip to tip. Of course, I could not take the eagle because it was impossible to keep it in captivity. Instead, I took a few pictures of the child with his eagle and I gave him a dollar. The child reacted as if I had given him $100. His mom and dad were so happy that I almost cried in front of them. After we left the area, I felt the tears running down my cheek. It made me so sad to see the suffering of these people who lived there. Suddenly I understood the resolve of the "wetbacks" who risked their lives in pursuit of the American dream.

Finally I was getting closer to the border between Mexico and the United States, so I expected a constant flow of people. This non-stop traffic of people and vehicles presents a problem that has existed since the last century. The flow may reduce or increase at any time, and there appears to be no way to stop it. It is well known that the illegal flow starts down in South America. The illegal activity is continuous, and one sees it at the border. The border between the U.S. and Mexico is quite large. It runs east to west from the cities of Brownsville, Texas, to Matamoros, Tamaulipas. There is also a border crossing in San Diego, California, and Tijuana, Baja California. The border runs along the Rio Bravo, which is called Rio Grande in the U.S. There are also large portions between the Sonora and Chihuahua deserts. In that stretch runs the Colorado River.

My goal was to cross the border at the city of Juarez. To my great surprise, the customs officers did not give me much trouble. They had a far greater interest in the Mexicans leaving the country and foreigners entering it. The biggest problem was, and will remain, the actual crossing of the border. The amount of different kind of vehicles, cars, buses, people walking, and other means of transportation was incredible. At any time of day or night, a stream of people and vehicles of all kinds crossed the border both ways. If the reader ever plans to cross this border, then he must take the following precautions: (1) he must have patience; (2) his vehicle engine may overheat, and (3) he must pass the border with a full tank of gasoline. Be prepared because at the border anything could happen. As I said before, the Mexican customs police have a greater interest in Mexicans leaving the country and foreigners entering the country. For the American border patrol, it is the other way. They are interested in illegal entering the United States and US citizens leaving the United States. Contraband at the border has existed for decades and will never stop. The articles of interest for both sides vary according to the market demand. I have crossed the border several times, and I always feel extremely nervous because one cannot predict what is going to happen. Entering the United States is not as easy as one might think. The U.S. Border Patrol is fortified with the most modern equipment imaginable. As an American citizen, I did not have much trouble re-entering the United States.

When I arrived in El Paso, Texas, we were hoping to eat a nutritious "hamburger." We went to McDonald's and ate "till we dropped." We left El Paso, Texas, in excellent spirits, but when we were getting near Las Cruces, New Mexico, both my wife and I experienced a bout of diarrhea so intense that we thought we were going to die. Later, I told Nora: "This is Geronimo's revenge." We had gotten so used to eating Mexican food that our stomachs began to reject American food. We went to a hospital and got the right treatment to stop the cursed diarrhea.

From Las Cruces to Albuquerque is about four hours, but, as I was pulling a trailer, it would take more time. My goal was to go to a hotel to relax and then unhook the trailer from my truck. The plates of my truck were still the original 1968 Puerto Rico plates. My driver's license was also from Puerto Rico but was not expired. Therefore, it was a priority to get New Mexico license plates and a valid New Mexico driver's license. Several days later, I went to the University of New Mexico to do the paperwork required to enroll in my new semester courses. That was not much of a problem as I had been speaking, writing, and reading English since I was a kid.

My other goal was to find a home or an apartment. The university had an office that helped students and their families find places to stay, and since I had money, it was easy to get an apartment that we liked. I found one that was well-located, not far from the university. It was also close to the entrance of the Sandia Air Force Base. Once we were installed in the new place, my mission was to pursue my post-graduate degree. Luckily, the Physics Department was combined with the Astronomy Department. I had already taught astronomy at the University of Puerto Rico; therefore, I was familiar with everything that had to do with astronomy.

As I mentioned earlier, when I was a professor of physics at the University of Puerto Rico, they were inaugurating the Arecibo Ionospheric Observatory. This observatory was "the new kid on the block." In 1960, Arecibo was still a small city. The observatory was several miles from the city in the middle of nowhere. I was there to check the facilities and then decide if I was going to do any research there. At that time, the highest priority of the observatory was to see if there was life and intelligent beings in space. The observatory was actually a gigantic ear that was listening to the universe for any indication of life in outer space. I was not interested at that time in that type of research. In those days, traveling from Rio Piedras to Arecibo was a journey of several hours. To get to the Arecibo Observatory, the road was narrow and difficult. It was, to say the least, an unpleasant trip. This observatory was inaugurated in 1963 and is still in used today.

I should point out that William Gordon, an engineer from Cornell University was responsible for the building the world's largest dish antenna in use by this observatory. He really started this project around 1958 when I was a Physics student at the University of Puerto Rico. Initially the idea was to study the Physics of the Upper Atmosphere due to the fact that missiles must travel through this region of space. The Advance Research Project Agency (ARPA) sponsored the project. The dish antenna is 1,000 feet in diameter and its big weight is really supported by the hollow natural crater in that region of Puerto Rico. I believed that Cornell University with funds from the USA Air Force administered the observatory for many years. I understand that as of this writing Ana G. Mendez University replaces Cornell University in the administration of the site.

Returning to my story of my life in Albuquerque, New Mexico, my wife complained that she felt awfully lonesome because we did not know any Puerto Ricans in Albuquerque. Also, she did not know where to get the ingredients for preparing Puerto Rican dishes. I did not know anyone either. I went to the Graduate Student Office and asked them to give me three names of Puerto Ricans living in Albuquerque. They gave me the three names, and among them was one who had been studying for a doctorate in Spanish Literature. That is how I met Hector and his wife who was of Mexican descent. Hector was a veteran of the Air Force and, by coincidence, he knew my brother Manuel who was also a member of the Air Force. Both of them had met in Tripoli, Libya, while both were stationed there. Hector then introduced me to other Puerto Ricans who were living in Albuquerque. All the Puerto Ricans who were stationed at the Air Force base eventually became my friends. My wife also became friends with the group, especially the wives. The Puerto Ricans gathered every Saturday in different houses. We rotated the meeting place. So after a few months, my wife and I felt at home. Since many are not familiar with New Mexico, I think it will be an excellent idea to learn more about this "land of enchantment."

New Mexico is a state in the southwestern part of the United States, and its capital is Santa Fe. It is bordered to the north by the state of Colorado, to the northeast by the state of Oklahoma, to the east and southeast by the state of Texas, to the southwest by the states of Chihuahua and Sonora (Mexico), and to the west by the state of Arizona and northeast

Utah. There is an area that is called the Four Corner area because that is where four states meet.

At the time of the conquest of Mexico by the Spanish Conquistadors, the territory included all states north of the Rio Bravo, i.e., what today is California, Texas, Arizona, New Mexico, Nevada, Utah and Colorado. Mexico was called the New Spain. Later on, the Spaniards decided to break it into two territories. Santa Fe became the capital of the northern territory, and Mexico City was the capital of the southern territory. The dividing line between them was the Rio Bravo.

In 1912, New Mexico was accepted into the American union. For hundreds of years, three cultures had settled this part of the country: the Indians, the Mexicans and the Spanish. When Sam Houston decided to seek independence for Texas, there was a conflict between the USA and Mexico. General Santa Ana, who was the leader of Mexico during that time, fought against Sam Houston. Santa Ana lost the war, and Texas became a state. The rest of the territory was in the hands of the American Indians. Once New Mexico became a state, the three cultures, plus new American settlers, merged into a new group of people adapting themselves to live in the new country that formed those states.

New Mexico is the state with the largest Hispanic and Indian populations in the United States. The elements of its history and geography make this state a different mix from the rest of the United States. The landscape and climate, traditional adobe architecture, family traditions, cooking and other cultural events have attracted artists, writers, and intellectuals over recent decades. Of course, we cannot ignore the hippies who have also contributed to the culture and history of the state. Unfortunately, I cannot explain everything because it would take several books just to discuss all the details of this famous American region.

Gradually I got to know the new environment around me much better. I was lucky that I was given an office in the Department of Physics, which meant that I could organize my own personal space. I had a place to study where nobody bothered me, and, on top of that, I had all the services that the university offered to its graduate students. As if that were not enough, they allocated a desk within the university main library so that when I was doing research in the main library, I had everything on hand along with a desk on the top floor of the library. There, I could spend long hours studying, and nobody bothered me. All these fabulous accommodations helped me to further my knowledge.

During the next two years from 1972 to 1974, I did extensive studies, not only in physics but also in astronomy. In addition, the head of the Physics Department offered me a position as a teaching assistant. They paid me some money for doing that job. It was not much, but it helped pay the rent on my house. Yes, I moved from my apartment to a house in a prestigious suburb. I no longer had to stay crammed into an apartment. I lived near the house of one of my professors and saw him several times walking his dog. Behind my house, was the Sandia Mountains. My Puerto Rican friends always called me when there was a party. During the holidays, we always roasted a pig and had all the delicacies of the Puerto Rican cuisine. Also, there was a Christmas party on the 24th of December, plus another party during New Year's Eve. We were starting to enjoy the holidays just as if we were in Puerto Rico.

I was writing a doctoral thesis in Physics of the Upper Atmosphere. I was studying the origin of the phenomenon that is commonly known as the "Aurora Borealis" or Northern Lights. The phenomenon is most common at the poles of the earth, i.e., at the North Pole or the South Pole. This phenomenon can occur in other parts of the world for short periods of time. In the northern hemisphere, it is known as the Aurora Borealis, and in the southern hemisphere, it is known as the Aurora Australis. "Aurora" comes from the Roman goddess of sunset, and "Borealis" comes from the Greek word "boreas" which means north. In Europe, it often appears on the horizon with a reddish hue, as if the sun emerged from an unusual direction. My partner who lived in Alaska was in charge of taking photographs from January to March when the phenomenon could be seen much better. I was commissioned to do the analysis of the photographs. Sometimes a NASA plane took photographs which were sent to my lab for analysis. There is a branch of science called "photogrammetric." It is essentially studying a photograph that is two-dimensional then transforming it into a photograph of a three-dimensional solid. In that way, you can make the correct interpretation of what you are watching. This involves some mathematical transformations which I will not discuss here because they are beyond the scope of this book. There are very few people who mastered this technique, and I became one of them. Although I cannot explain the details, I can give an example. After a plane flew over Cuba taking thousands of pictures, the photographs were analyzed by experts in the field. That gave President Kennedy definite evidence that the Russians were erecting rocket launchers in Cuba, triggering the Cuban-Missile Crisis. And the rest is history. The photogrammetric process is essential for military applications. Every day, various aircraft and satellites take pictures of enemy territory. After studying the photographs, one can understand what happened on the ground. Because of my experience in many subjects within the field of physics, one of my teachers made the necessary contacts for me to work at Los Alamos Scientific Laboratories. Therefore, in the summer of 1974, I started working as a graduate student in these historic laboratories.

LOS ALAMOS LABORATORY IN NEW MEXICO

I'd like to give a brief history of this famous laboratory. In early 1940, work began on the development of nuclear energy. German, Russian and American scientists were studying the process known as "atomic fission," the process required to develop an atomic bomb.

The laboratories are located a few hours from Santa Fe, and to reach them, one had to travel along a narrow road (at least at that time) with many cliffs on one side of the road. Situated in the middle of a forest, the site is quite cut off from the rest of the surrounding towns. I had made the necessary arrangements so my wife and two children could stay with me. The only apartments available were some that were built in 1940. They were made of wood and had two floors. My apartment was on the first floor and was fully furnished except for personal belongings such as my kids' toys. It was a comfortable apartment except for the noise from above. The neighbor upstairs walked a lot and the ceiling shook under his weight. But after I complained to the administration, the noise was reduced to an acceptable level.

I cannot discuss the kind of work that I did there. It is well known that these research laboratories were primarily involved in various branches of physics. Many of the applications were intended for domestic use; other applications were for military use. For example, they were developing a vehicle that used no gasoline. The idea was to create a car that used water as the primary energy source. This was called a peaceful application. There were other laboratories in other parts of the country working on the same idea. The spin-off is the hybrid cars that we see today on the market. The vehicle runs on batteries part of the time and gas the rest of the time while the batteries are charging. The lab equipment was quite advanced. It was there that I began to specialize in computers and computer programming languages.

I learned to write programs in FORTRAN, Pascal, COBOL, and a language called "machine language." There I also met many physicists and mathematicians who had won the Nobel Prize in Physics and Mathematics.

I enjoyed socializing with scientists, and many of them invited me and my wife to their homes for dinner. On those weekends when there were no parties, we watched TV or went sightseeing in the area. The nearest town was Santa Fe, but driving at night from Santa Fe to Los Alamos was considered suicidal due to the narrow road.

I learned that the suicide rate was higher there than expected. I never understood the reason for so many suicides. Perhaps when one lives in an isolated place, one loses touch with reality. My partner who took the photographs in Alaska told me that the suicide rate there is also unusually high. Winters in Alaska are extremely harsh. The nights last up to 20 hours with only four hours of sunlight. In the summer, it is the opposite, with the daylight lasting up to 20 hours and the nighttime lasting only four hours. In Alaska, alcohol use was accompanied by the rise of suicides. On Indian reservations in New Mexico, alcohol and suicide also occur together. On Apache, Jicarillas and other Indian reservations, alcohol and suicide account for the most deaths. My opinion is that idleness leads to alcohol, and alcohol leads to suicide in these groups. There are other theories. I do not know which views other people may have on the real cause of suicides.

In late summer of 1974, I finished my work in Los Alamos. On my last day, I was given a farewell fit for a king. I got some gifts and my boss asked me to come again the following year. Although I enjoyed working there, living in such a remote place is certainly not for me. I never went back because in the summer of 1975 something remarkable happened. However, what I learned in Los Alamos would help me in my future. That was the objective that my mentor was seeking.

APACHE INDIANS AND OTHER TRIBES

When I was a boy, I loved going to watch cowboy movies, and the majority of these films depicted Indians as the villains. Now I was living in the region of the Apache Indians. Luckily, one of my fellow students named Jesus Algodones was a member of the Apaches Mezcaleros tribe. I asked him to take me to his reservation so that I could learn about their culture. He told me that to go to his house we needed a 4-wheel drive vehicle. I told him that I owned a 4x4 GMC Jimmy. So one day, he took me to a cultural festival where I was welcomed as a guest of his tribe. I was amazed at the number of tribes that lived in the region. At the point where four states met, there was one of the most popular Indian reservation areas. I think the following tribes still live in that area—Navajo, Jicarillas-Apache, Comanche, Pueblo, Zuni, Chiricahua, Mimbreno-Apache, Mescalero Apache, Jocome, Janus, and Suma. In this book, I have no time to discuss them all. I will only consider those that I visited with Jesus.

The Apache Indians live in the states of Arizona, New Mexico, and Texas. They still retain their own languages which are divided into two—Apache-East, and Apache-West. Rather like French and Spanish, the languages are related, but they cannot understand each other. There is a third language spoken by the Navajo and Apache-west. The things that most caught my attention were the use of alcohol and poverty. It is heartbreaking to remember how Americans snatched their lands and confined them to the four corners of four states. This pain passes from generation to generation.

The Navajo Indians have a single language that is spoken only among themselves. They are probably the richest Indians of all the tribes. They control large businesses in the area. They have enough money and live comfortably. At the time that I visited, there were more than 150,000 Navajos living on the reservation. Their leaders are respected by most of the tribe, and they try not to involve themselves in the affairs of the white man. There are probably more Navajos than any other tribe in the United States.

The Pueblo Indians are located in New Mexico and Arizona. They are subdivided into several groups of Indians who have a common ancestry. Each group has its own dialect. I visited the Acoma, Cochiti, Laguna, San Felipe, Santa Ana, Santo Domingo, and Zia. But there are other groups as well: Isleta, Jemez, Nambe, Picuris, Pojoaque, Sandia, San Ildefonso, San Juan, Santa Clara, Taos, Tesuque, and Tigua/Ysleta. It is fascinating to note that the Spanish colonists were in this area before the Americans; therefore, all these tribes and groups of Indians have Spanish names and many still speak Spanish, making them multilingual. I spoke to them in Spanish, and many of them understood me. The surprising thing was that sometimes I felt as if I were back in Teotihuacan, Mexico. Many of these groups were descendants of the Aztec Indians, who lived there before the Spaniards arrived. There are two groups known as the Hopi and Uto-Aztecs and their language is related to the Aztecs who spoke Nahuatl. The sad thing is that, for the ordinary American, these tribes and groups are invisible. If one studies the history of the relations between the tribes and the U.S. government, one will see the injustices committed against the Indians. My friend told me that not only did they take their land and kill women and children, but relocated them in reservations. Even today there is no easy solution to the problem. For this reason, I call them "the invisible people." As I said earlier, Indians are portrayed in movies as the villains. This is understandable as the history of the wars between the Indian tribes and the new settlers was written from the point of view of the victors—the white people.

STUDY OF THE EARTH ATMOSPHERE

After my return from Los Alamos, I continued working on my doctoral dissertation in Astronomy. As mentioned earlier, my doctoral dissertation was based on the phenomenon of the Aurora Borealis. It was then that my thesis advisor approached me and told me that the National Institute of Health (NIH) was looking for Hispanic or Latino graduate students. They had launched a new program where one could become a medical doctor in two years if one were willing to work in El Paso, Texas, in partnership with the School of Medicine at the University of Guadalajara in Mexico. The offer was tempting, but it meant that I would have had to shuttle back and forth between the two cities for a short time. I thanked my thesis advisor and told him that I preferred to continue working on a doctoral thesis sponsored by NASA.

It seems that my advisor brought me miserable luck because, in the 1974-75 timeframe, the infamous gas crisis occurred, nearly paralyzing the country's economy. Despite the energy crisis, I kept working on the analysis of the photographs that were taken in Alaska and, using physics principles, I developed a theory to explain the phenomenon of the Aurora Borealis. I had already completed all Ph.D. requirements at the University of New Mexico.

Around March of 1975, my advisor said that he had already made arrangements for me to go to work in the summer of 1975 at the Marshall Space Flight Center (MSFC) in Huntsville, Alabama. By coincidence, MSFC had a lab that was also studying the Physics of the Upper Atmosphere, and that would give me a chance to work and collaborate with NASA scientists on studies of the Earth's atmosphere. Skipping the technical details, I need to consider the issue of High Atmospheric Physics because this new adventure in Huntsville, Alabama, would have a tremendous impact on my future life.

The Earth's atmosphere has several layers as follows:

The **troposphere** reaches an upper limit (tropopause) of nine kilometers in height at the poles and 18 kilometers at the Equator. It produces significant vertical and horizontal movements of air masses (wind) and relative abundance of water there. It is the region of cloud and other weather phenomena: rain, wind, temperature changes and the layer of most interest to ecologists. The temperature decreases as one goes higher.

The **stratosphere** begins at the tropopause and reaches an upper limit (stratopause) at 50 kilometers altitude. The temperature changes and is around zero degrees Celsius in the stratopause. There is almost no movement of air in a vertical direction, but the horizontal winds frequently reach 200 kilometers per hour, which facilitate the rapid spread across the globe of any material that reaches the stratosphere. For example, this is what happens to the ozone-depleting CFCs. In this part of the atmosphere, between 30 and 50 kilometers, the ozone layer is beneficial because it absorbs harmful solar radiation.

The **Mesosphere** which extends between 50 and 80 kilometers in altitude contains only about 0.1% of the total mass of air. It is important for the ionization and chemical reactions that occur therein. The decrease in temperature, combined with low air density in the mesosphere, determine the evolution of turbulence and atmospheric waves, setting in motion high spatial and temporal scales. The mesosphere is the region where the space shuttle, returning to Earth, begins to experience the wind resistance that starts its deceleration.

The **Ionosphere** extends from a height of about 80 kilometers above the Earth's surface up to 640 kilometers or more. At this distance, the air is very thin. When particles in the atmosphere undergo ionization due to ultraviolet radiation, they tend to stay ionized due to minor collisions between the ions. The ionosphere has a big influence on the propagation of radio signals. A portion of the energy radiated by a transmitter into the ionosphere is absorbed by the ionized air, and another is refracted or deflected again to the surface of the Earth. This latter effect allows the reception of radio signals at distances much greater than would be possible with waves traveling through the earth's surface. The region that lies beyond the ionosphere is called the **exosphere** and extends to 9,600 kilometers, which is the outer boundary of the atmosphere. Extending further out is the **magnetosphere**, the space around the Earth where the planet's magnetic field dominates the interplanetary magnetic field.

At that time, no one knew for sure what caused the Aurora Borealis. In order to achieve the necessary research, I had to go to Huntsville, Alabama. I was working in a laboratory in Huntsville in the summer of 1975. My wife and children remained in Albuquerque, so I was living alone. At that time, Huntsville was a small town near the border of Tennessee. I went in my truck to explore the area during the weekends. I visited the home of Elvis Presley. He was living there at the time he died. The house then became a museum and one could spend two to three days looking at all the Elvis memorabilia. There was also a place in Memphis called "the painters' alley," where many hippies and intellectuals used to hang out. I also saw live performances of music known as "country music." It was there that I learned to appreciate that kind of music and the famous artists who played it. I also went to a place that was called "the Grand Ole Opry." It was a great big house with a huge stage where country music was played exclusively. It was there that I saw a live show with Johnny Cash. I spend my spare time touring the area, especially the southern part of Tennessee.

"Redstone Arsenal," the military base, was large and covered several square miles. Within the base there were many laboratories. The "Marshall Space Flight Center" (MSFC) was also located there. I remained there until the end of July 1975, and then I went back to Albuquerque. If you visit this part of the country I encouraged you to tour the area, especially the Marshall Space Flight Center museum. They have many spacecraft on display and the history of MSFC. This center was responsible for the development of the propulsion rockets used in the Apollo and Space Shuttle Program.

MY ENCOUNTER WITH WERNER VON BRAUN

When I was in Huntsville, Alabama, one day I received a delightful surprise. The Universidad Autonoma de Mexico (UNAM) had found all of my documents, and I could pick up my Doctor of Physics diploma. They could not send it by mail, so I had to fly to Mexico City to verify my credentials in person. Luckily, one of my teachers in Mexico, Dr. Andrade, offered me accommodations in his house. In fact, when I arrived in Mexico, Dr. Andrade was at the airport and drove me to his apartment where he lived alone. Then he handed me the keys to his apartment and told me that he was going to stay at his parents' house. It was the first time I ever slept on a water bed. This type of bed was very popular at that time, but I did not like it. Every time I moved, it felt like a warship was moving over the waves. I spent a week eating Mexican food and watching some variety shows in some Mexican theaters. I went to the University, and, after fulfilling all the requirements, I was given the diploma. Both Dr. Andrade and I got drunk to celebrate that victory.

The next day, I took a bus for my return trip to Albuquerque. Traveling on a bus in Mexico is quite an experience. This was not one of those comfortable, air-conditioned buses used by tourists, but rather one used by the Mexican working class. You got to see beautiful scenery while rubbing elbows with the lower classes. As I said earlier, for Mexicans, I was a Yucateco, from the Yucatan peninsula. In the United States, a bus has a limited occupancy, and everyone must be seated. In Mexico, there are no occupancy limitations. The capacity is whatever number can be accommodated on the bus—seated, standing, or on the roof. Add to that luggage, purses, bags, and even animals. To signal the driver to stop, the passengers yelled to the driver, banged on the ceiling, or pulled a string (if the string was working). Then the problem was to get off through one of the side doors or the one front door next to the driver. To make matters worse, in Mexico there were no tarred roads, and when it rained, it was another experience. My wife and kids did not like traveling on Mexican public buses. I, on the other hand, liked it, because it gave me the opportunity to observe Mexican people and their customs. Maybe I should have been an anthropologist rather than a physicist.

My senior year at the University of New Mexico was not easy because I had already completed the courses and was defending a doctoral thesis. Some students complete a thesis in a year while others spend four years and still have not finished. My friend George had already spent four years, Fred had spent three, and I was entering my second year. Chang, my Japanese friend, got tired and went to work at Jet Propulsion Laboratories in California, earning good money.

I went to my thesis advisor and asked him how much longer I had to spend defending my thesis, and he responded, "four years." I explained to him that I was also a professor of physics at the University of Puerto Rico; I had already been out of the country for five years, and I wanted to return to my faculty chair at UPR by August of 1976. My last semester at Albuquerque ended in May. Then the summer session started. My savings were now gone, and I suspected that I could not stay in New Mexico much longer. He told me he would see what he could do to help me financially. (The thesis advisor did not know I already had a doctorate degree from UNAM.) Then one day my thesis professor called me into his office to give me the good news that I had gotten a job to work as a graduate student at NASA during the summer of 1976. Well that solved my problem. In addition, I was getting closer to the eastern United States. My truck and I were tired from traveling for so long.

While living in Huntsville, I observed that the great power once enjoyed by the Ku Klux Klan (KKK) was fading, but still alive. Alabama was the cradle of the KKK, and their presence was still evident in Huntsville. I saw signs saying "whites only." There were places you could not enter if you were black. I saw professional black people such as lawyers and doctors barred from these places. Public transportation such as buses and trains were still segregated. As I was white, I had no problems. As I spoke English with a German accent, people treated me with respect. The governor of Alabama had been a tireless advocate for the white race. In fact, he was seriously wounded for being an outspoken advocate for racial discrimination. To better understand the problems that I faced in Alabama, I first need to set the stage.

In 1963, things had reached the boiling point in Alabama. The struggle over racial discrimination was increasingly volatile, and word reached the nation's largest newspapers. George Wallace used the issue of segregation as a political tool to get elected again. So on June 11, 1963, Governor George Wallace stood in front of the Foster Auditorium at the University of Alabama to block black students from entering.

During his 1968 campaign, the issue of segregation came to a climax. During one of his speeches, a shot rang out, and the governor fell to the ground, critically wounded, but doctors managed to save his life. The governor was paralyzed from the waist down—the .38 caliber bullet remaining lodged in his spine. However, the shooter was not a black man, but a young white man about twenty years old. Wallace was not the only victim of violence. That decade also saw the assassinations of President John F. Kennedy, Robert F. Kennedy, and Martin Luther King.

Wherever I went, I had always faced all kinds of obstacles—at the University of Puerto Rico, Columbia University, the National Autonomous University of Mexico, and now in Huntsville I expected the worst. I never got scared; I just kept pushing forward until I reached my goal.

In May of 1976, I came back to Huntsville, Alabama, but this time I was going to work and earn more money. I received orders to report to the offices of Dr. Werner Von Braun because he was the director of the Marshall Flight Space

Center. At that time, my English was what they called "broken English." When I entered Dr. Von Braun's office for the first time, he asked me if I had changed my name. I told him "yes," and explained that on my birth certificate, my name appeared as "Juan Rivera Diaz" but that my father registered me in the Catholic Church, and the baptismal certificate says "Juan Pedro Rivera." I told him that some people knew me as "Pedro Juan," others as "Pedro," and still others as "Juan." My mom and my sisters and brothers called me "Perin" which is my nickname. Then he asked me if I was German. When I told him I was Puerto Rican, he told me that I didn't look Puerto Rican and that I could pass as a German in Huntsville because I spoke English with a German accent. He said if anyone asked me if I was German, I should answer in the affirmative. Then he told me to go and report to Mr. X, who would examine my papers to make sure that everything was in order.

Being by myself in Huntsville, with my family still in Albuquerque, I found myself associating with the Germans who made up a large portion of the people in Von Braun's office. Almost every day, I got to interact with the Germans scientists and engineers during lunch time. I must confess that I learned a lot from the German coworkers. My plan at the time was to continue working in Huntsville and then see what would happen at the end of the summer.

THE SCIENCE OF ROCKETS AND MISSILES

It was then that I became interested in studying space science on my own, especially rockets and missiles. NASA's library is one of the best in the world, and I had access to everything that I wanted. I started by studying the works of Werner Von Braun when he was in the service of Hitler's Germany. I studied in depth everything related to rocketry, the development of jet engines, the chemistry and physics of the fuels used in rockets, and the different types of rockets and missiles that the Germans had developed during Hitler's regime.

As a physicist, I was able to understand all the principles of rocket science. Basically, they are based on Newton's laws of mechanics, specially the third law which is known as the law of action and reaction. For those not familiar with this law, I will try to explain it in simple language. The law says that "for every action there is a reaction equal in magnitude and opposite in direction." So, for example, if you fill a balloon with air and then release it, the balloon flies in one direction and the air goes in the opposite direction.

The next day at a ball park, I made a simple rocket using two plastic one-liter soda bottles, like those sold in supermarkets. I cut the two bottles in half and then joined the two conical parts forming a tube, ending in two cones. At one end; I inserted a cork to contain the water. Then I filled the tube with water. On the other open end, I pumped in air using a bicycle pump, so the water inside the pipe was under pressure. When the pressure reached a certain level, the cork was released and up flew the rocket.

What I had made was a water rocket. I cannot explain all the details in this book. Basically, the bottle rocket is driven by water under pressure. A water rocket is actually easy to build, and I had a lot of fun doing it. Water rockets operate under the same principles as the rockets used by NASA and the military. The idea is to create pressure within the bottle by introducing air. I used a bicycle pump to pressurize the water. A week later, I went to Dr. Von Braun to show him my water rocket, and he laughed for a long time. Then he gave me some ideas how to make rocket fuel using a recipe whose ingredients were easily available in the supermarket. At that time I noticed that his health was declining. After that day, I never saw him again.

Every day I continued perfecting my water rocket, but one day I decided to see if I could make a rocket using a better fuel. Of course, I cannot reveal what I did, but a rocket is merely a vehicle or spacecraft that generates its thrust through the rapid expulsion reaction of combustion gases. Therefore, there has to be something to raise the pressure of the gases, and the combustion has to be controlled; otherwise the rocket will explode. A missile really is another type of rocket having a predetermined mission. Missiles and rockets usually have to have a navigational system to drive them to its intended target.

So in a short period, I was able to learn the basics of rocketry and missiles. This shows that if you set a goal you can reach it. It just takes a lot of sacrifices and perseverance. The Germans always gave me ideas or tricks that were not found in books. There are different types of fuel. Some fuels are solid; others are liquid. I also learned that there are different types of rockets, especially those that were developed at a later date. Rocket science is quite complicated because everything has to be carefully controlled; otherwise the rocket will explode with pieces flying in all directions. There are rockets which are guided to the target by a certain technology. In short, rockets, military missiles, jet engines, and other variations use the same principle of physics.

During the time that I was at the Redstone Arsenal in Huntsville, Alabama, I was involved in rocket science in one way or another, and that would be the determining factor in my career as a scientist for the next 30 years of my life. That will become apparent as one continues to read this book.

So to summarize, a rocket is really a missile, whose range can vary at will. In ancient warfare, guns were used with a range of perhaps a mile, but then they developed missiles whose range was several miles. If one could continue to increase the range of a rocket to fly farther, there would be a point where the range of the projectile is the same as the circumference of the earth. In this case, you could shoot a rocket, and sometime later, after that rocket goes around the world, it could come back to its point of origin. The idea of a rocket that can go into orbit around the earth is the next logical step. The truth was that I did not realize that I was becoming an expert in rocket science. My life had taken a different turn. Everything I learned in my career in physics from the University of Puerto Rico, the National Autonomous University of Mexico, and the University of New Mexico—all that knowledge converged into my position as a NASA scientist.

The precursor of the rocket is based on the discovery of gunpowder by Chinese alchemists. This means that the use of rockets is certainly not a new science. When I was a kid, I watched fireworks during the festivities in my hometown. So at one point, I was experimenting with gunpowder as a fuel for a rocket. The major issues in the study of rockets relate to nozzle velocity and range. The nozzle velocity and the aim problems are easy to solve. One of the biggest problems is that of stability. You need to know what a vector is and the role it plays in rocketry. Because you need to be knowledgeable in higher mathematics, the rest of this research is beyond the scope of this book.

Russian and German scientists had made great progress in this field. The research work of Robert Goddard also helped in the development of this science. It is common knowledge that oxygen and hydrogen in its liquid form is an excellent fuel to propel a rocket. Because I understood German, I was able to read the essay of German scientist Hermann Oberth. What Oberth wrote was a major contribution to rocket science. The German scientists did a lot of work from 1927 to 1931 while experimenting with propulsion systems using liquid oxygen and gasoline. Russian scientists, under the leadership of Valentin Glushko, introduced certain regeneration techniques including chiller, hypergolic ignition, and fuel injector designs that added internal mixers and injectors that provided secondary propulsion.

Werner von Braun, under the leadership of Hitler, developed a series of rockets called V-1 and V-2 which originally belonged to the A-series of rockets. The initial name of the V-2 rocket was A4. Many of the V-2 rockets fell on the city of London. Those rockets had a range of up to 300 kilometers and carried a warhead of 1,000 kilograms of Amatol. They had turbine pumps, inertial guidance, and other capabilities. Thousands were launched against Allied nations, mainly England, France, and Belgium, but they really did not affect the outcome of the Second World War due to sabotage among the workers who built them. Hitler's mistake was in using the labor of prisoners of war. There is no doubt that the technology worked well. They only needed to correct the defects.

After the war, the United States captured Werner Von Braun and his team of scientists. Russia captured the rest of the German scientists. Both countries also captured the rockets that had not yet been fired along with all the rocket parts. The U.S. and Russia also acquired all the schematic diagrams and technical papers. They called this mission "The Paperclip Operation." Many things happened from the end of the war until the 1950s. When the Russians fired the first earth satellite, the V-2 evolved into what later was called the Redstone Rocket that was developed in Huntsville, Alabama. This rocket was used in the early stages of the NASA space program.

On the other hand, the Russians, using German technology under the leadership of Sergei Korolev, took the lead in the space race between the U.S. and Russia. The V-2 was the basis for the development of new Russian rockets that were called R-1, R-2, R-3, and so on. The Russians also developed new propulsion engines under the leadership of Glushko. Under the guidance of the Russian Aleksei Isaev, they developed what they called Intercontinental Ballistic Missiles (ICBMs). The Russians also used a distribution of rocket propellants around the tank which gave more power to put a satellite into orbit and put a man in space. Thus began the space race between Russia and the United States. The science of missiles is the same as rocket science. The biggest difference is that missiles usually have their applications on the battlefield. There are many types of missiles and various types of platforms. The launch pad can be the ground, a warship, an airplane, or a helicopter. Missiles can be classified by the way that they are launched—ground-to-air, air-to-air, or sea-to-air. There are many companies dedicated to the development of different types of missiles, and usually the best customers are the governments of different countries.

KENNEDY SPACE CENTER IN FLORIDA

My job at NASA had changed from graduate student to researcher. One day, I went to the human resources office and asked them what I needed to do to transfer from Huntsville to the Kennedy Space Center in Florida. The woman, who was helping me, gave me the phone numbers of the NASA Cape Canaveral Human Resources Office. She explained to me that while she was doing the paperwork from her end, I could call the Kennedy Space Center to see what paperwork was needed from their end. I made several calls to Cape Canaveral to find out the requirements for the transfer. Meanwhile, I continued working in the lab, and sometimes I went to the library to continue studying the science of rocketry.

I received a call from the human resources department at the Kennedy Space Center, and I was told that I could not get a transfer because there were no vacancies for scientists. What they needed were engineers, and being Hispanic gave me a better chance to get a vacant spot as an engineer. I was told that with my credentials, I qualified as an Aerospace Engineer. On the other hand, the woman from the HR office in Huntsville told me they had done the paperwork for me to stay permanently as a scientist in Huntsville. I would be starting with the salary scale of a GS-9 (because it was what they had at the time), but after six months, I would move to a GS-11, and after a year, I would go up to a GS-12, provided I had performed a satisfactory job during the first 6 months. If I did not meet the requirements after a probationary period, I would be out of work. I have always had faith in myself and my intellectual capabilities, and I knew that I had a secure job with NASA, even if I had to start at the bottom.

Earlier, I had advised the readers of this book that when climbing a ladder, one has to start with the first step. If one tries to climb higher, one is more likely to slip and fall, with possibly injurious results. I knew I had to climb the ladder one step at a time. What NASA needed at that time in Kennedy Space Center (KSC) were many engineers. The mere fact of belonging to a minority—as they considered Blacks and Hispanics—turned out to be a plus instead of a negative. The administrations of John F. Kennedy and Marshall Space Flight Center were committed to obey the laws that protected the rights of minorities. This meant that the political situation was leaning in favor of minorities. So I could achieve the KSC-MSFC transfer if I was willing to start at the low end—GS-7. The salary remained the same because it was what they called a lateral transfer. The biggest problem we had in the transition from KSC-MSFC was that I still had to solve several problems of logistics. First, I had to put my status as a student at the University of New Mexico on hold because they gave me a year to come back if I wanted to defend my thesis on the Aurora Borealis. Second, I had to decide whether or not to return to UPR as my professorship was also on hold for a year. However, there was a possibility of converting my time at NASA to a sort of license without pay. There was already a precedent of a university professor serving in another government agency to fulfill its services in that other agency and then return to the chair without losing rights or pension. Third, I had rented and paid the rent on the house in New Mexico. My wife and children were still living in New Mexico, and the children were leaving school. The first two problems could wait a year, but moving out of Albuquerque was a priority.

My older brother, Roberto, lived in Apopka, Florida, and offered me shelter at his house until I found a home. My truck was with me, but the trailer was in a private garage in Albuquerque. The problem was that I needed two trucks to move my entire household to Florida. Through the years, I had bought a lot of souvenirs on all my trips. I had a collection of antique radios, and various objects in onyx and silver that I had bought in Mexico. I also bought a jukebox in Mexico. A coin had to be deposited first, and the jukebox played songs on large 78 revolutions per minute records. Therefore, before heading to New Mexico, I had to pick up a driver in Huntsville.

This man was African-American and had become my friend. He also was an auto mechanic; therefore, he could help with any car problems on the road. Besides, I needed someone to help me pack all my belongings into the trucks. My "stuff" had grown to nearly double what it was when I left Miami, bound for Mexico.

"El Puma," as I called him affectionately, had never been in the southwestern United States and was amazed to discover new lands he had never seen in his life. To go to Albuquerque, we had to take route US-40. This road crossed through northern Alabama and Mississippi, and then crossed the states of Arkansas, Oklahoma, and Texas into Santa Fe and finally south to Albuquerque. When I lived in New Mexico, I never saw a black person. Therefore, my friend was amazed by everything he saw, but at the same time, he told me he was a little nervous because he had been a victim of whites. I told him not to worry because the Indians had also been victims of whites in New Mexico and there were still many Indians. We made the trip in 24 hours straight—three shifts of eight hours each until we arrived at my home in Albuquerque. I decided to buy airline tickets for my wife and children to fly to Florida to stay with my brother in Apopka until I arrived with the moving trucks. After my wife left, I made arrangements to load a truck that my friend would drive. I planned to pull the trailer with my truck. When all vehicles were fully loaded, I told El Puma to just follow me closely. I was going to lead the way at a speed of 50 miles per hour.

We left early in the morning from Albuquerque, stopping only for gas, food, and other needs. This time, we took another route, US-10. This route crosses the entire United States along the southern parts of the country. The road begins in Jacksonville, Florida, and ends in Los Angeles, California. At that time, it was a new road. Everything was going well until malice or greed took over El Puma. When we were crossing Mississippi, he suddenly vanished—disappeared with the U-HAUL truck and all my belongings. I stopped in New Orleans and went to the highway patrol for help. It took them several hours to find El Puma who was eating at a restaurant. The police asked me if I wanted him arrested. El Puma told me that he had gotten lost and did not know where I was. I never knew if it was the truth or just a convenient lie, but I told the police that I needed him, so it was better not to arrest him. After losing time over that emotional experience, there were no more hitches for the rest of the way back to Orlando. I rented a place to put my trailer and all my belongings. I paid El Puma, because despite what happened, I liked him anyway. I bought him a plane ticket and told him that one day I would see him in Huntsville, Alabama.

I started working at KSC in May of 1976. Although I was 38, people thought I was only about 28 (thanks to the right genes inherited from my mother who never looked her age). When I reported to the human resources office, I met Mrs. Merrilees, who told me she was going to be my supervisor until I found a permanent job within the center. That did not surprise me because it was customary for them to form a pool of 10 to 20 engineers. They could be assigned wherever they were needed. She told me that since I had scientific degrees, she would put me to work in the scientific laboratories.

The laboratory had physicists, chemists, biologists and experts in metallurgy, fluids, cryogenics—a combination of experts in all branches of science and technology. My boss was Dr. Gail who was from Scottboro, Alabama. He had a doctorate in theoretical physics, so he and I formed a team engaged in research. As he was the director of those laboratories, I found myself connected with the right person. I noted that some of the lab people displayed signs of racism because I was a Puerto Rican, but Dr. Gail became my protector and defender.

THE FIRST LAUNCH OF THE SPACE SHUTTLE

At that time, we were making a transition from the Apollo program to the Space Shuttle program. The last Apollo mission was the Apollo-Soyuz. To summarize, the goal was to have the two spacecraft dock in space. Then the crew from one vehicle would transfer to the other vehicle. The idea sounds simple, but it is a rather complicated task. First, they had to make adjustments to both vehicles traveling in the same orbit, chasing one another. On July 15, 1975, Alexey A. Leonov and Valery N. Dubasov were launched into space from the Baikonur Cosmodrome in Kazakhstan, Russia. That same day at 3:50 p.m. from "Launch Complex 39" in Kennedy Space Center, the last American counterpart was launched. Thomas P. Stafford, Vance D. Brand and Donald K. Slayton were the astronauts chosen for this mission. The orbit had a radius of 220 kilometers. One of the vehicles approached the other to make contact, and then the necessary adjustments were made for the crew of one vehicle to dock with the other. The mission was a success. It proved that the technology called "docking in space" was possible, and it also opened the doors to mend relations between Russia and the United States. This maneuver was especially critical for the success of future space missions. Unfortunately, no one knew that this was going to be the last flight of the Apollo program.

After the last Apollo mission, we started working on the preparations for the Space Shuttle's first flight. We started dismantling everything out of the Apollo program and started building all the tools needed to launch the first Space Shuttle. As this was a new mission, it was an intense and tremendous job. There were many unknowns. It was a new concept and everything was different. Therefore, my work in the laboratory had enormous importance.

What caught my attention was that every time I went to have lunch at the cafeteria, all I saw was gray hair. I thought I was in a nursing home. I did not know anyone. So my first job was finding someone of Latin blood. Luckily, one day I was walking down the hall and I heard someone speaking Spanish. I stopped and asked, "What is your name?" He answered, "My name is José González." "Where were you born?" I asked again. The answer was: "I am from El Paso, Texas." Then I said, "Well, you are my friend." Jose told me they had a few Mexicans, Cubans and two Puerto Ricans working at the center. This was the beginning of my relationship with Latinos. Then I met a Latino from New York. His name was Sam. Later on I met Juan Torres who became my friend for the rest of my life. One day, I was reading a newspaper that circulated within the NASA space centers. It was reported that NASA had to meet a quota. Ten percent of NASA engineers had to be members of a minority. As we Puerto Ricans qualified as a minority, I told the human resources person in charge that I had been a professor at UPR and that the Mayaguez campus had a remarkably strong engineering school. I was sure they could get many candidates from that campus. She told me she was going to discuss the idea with her boss. Well, about two or three weeks later, Human Resources sent a Latino (a Filipino named Manuel "Manny" Virata) to Mayaguez. In short, during the period 1976 to 1980, they hired 25 engineers from the UPR Mayaguez campus! I was extremely pleased to see that I was not the only Puerto Rican engineer. In fact, we Puerto Ricans, Mexicans, and Cubans also contributed to the development of the space program in the United States.

One day, I met a group of Puerto Rican engineers in the cafeteria. An Anglo engineer was offended because we were speaking in Spanish, and he complained to the director of the Space Center. At that time, the director was Robert Smith, and one day, he asked me to report to his office. Upon arrival, he told me that someone had complained that we were speaking Spanish in the cafeteria. I responded that lunch time was not part of the working time. We were talking among ourselves. We had not broken any rules. I asked him to please send an official statement from his office clarifying the issue. The memo should be addressed to all Space Center employees stating that "the language for all official conversations either by phone or in official meetings had to be in English." Meetings in the cafeteria were not to be considered official meetings.

To calm down the protesters, the Center director also hired the services of a language teacher to instruct us how to minimize our Spanish accents. Ironically, the language teacher was from Alabama and had a horrible Southern accent. What a joke to think that an accent can just be removed like turning a switch! The Puerto Rican group took the course to fulfill the requirement. The teacher from Alabama was extremely pleased. He earned a few bucks, but the answer to the problem was not entirely solved. The Anglos finally had to accept the fact that thousands of accents, including Southern, Yankee, Texas, and Latino all existed in the United States. This linguistic obstacle was the first that the Puerto Rican engineers had to overcome.

I agreed with the Center director that I and others had to reduce our accents and try to use the English language more fluently. I hired a trainer from Orlando, Florida, to correct my Latino accent. That way my conversational English would be more acceptable to the rest of my co-workers. Months later, my boss told me jokingly that I did not talk with an accent anymore. I kept quiet, but I had reached my goal.

BECOMING A NASA ENGINEER

Let's talk about engineering. There are different branches of engineering. Among the best known are civil, agricultural, mechanical, electrical, and others. Engineering is applied physics. That is, most physicists devote themselves to studying the basic principles of forces (civil engineering). In physics, we call it "static," and when forces cause motion, we call it "dynamic." Therefore, engineering students must complete the core courses in physics first and the necessary courses in mathematics. Physicists, by contrast, study all the basic principles of all branches of engineering. For example, I studied all the basic principles of statics, dynamics, electrostatics, electrodynamics, hydrostatics, hydrodynamics, magnetism, electricity, mechanical waves, sound waves, electromagnetic waves, atomic physics, nuclear physics, physical optics, electro-optics, and others. A trained physicist must be an expert in any of these disciplines. A physicist must also have a clear understanding of college algebra, plane and solid geometry, trigonometry, advanced algebra, calculus, ordinary differential equations, partial differential equations, Boolean algebra, logic, computer programming, and many others. As I had a complete understanding of all these subjects, it was easy for me to understand and solve all kinds of engineering problems. When I was working at different places, I showed my supervisors that I had the intellectual credentials and experience to deal with most engineering problems.

While working as a scientist at the NASA laboratories, one day my boss called me into his office. He said, "Look, Juan, the administration has decided to carry out a reorganization of NASA engineers here at the Kennedy Space Center. As more engineers are needed, we have decided to change your job title to an engineer. From now on, you will work as an electrical engineer or a computer engineer instead of a scientist. The advantage you have is that you know computer programming and also know about the electronics of a computer." I replied, "So that means I'm going to go to another group." He answered, "Starting next Monday you have to report to Mr. William who will be your immediate supervisor. Mr. William is a Division Chief, and you will be working for him directly. "

My co-worker scientists at the laboratories gave me a farewell party. Inside the Kennedy Space Center, no alcohol can be served. So they decided to take the party outside the Center. At the end of the farewell party, they told me that the doors were open if I ever wanted to return to work at the laboratories.

Those who have seen rocket launches on television have noticed that there is a spacious room with a large group of computers and engineers. These computers monitor and control all systems that have a rocket or the space shuttle or even a space plane. Since 1960, computers control and monitor all systems of a spacecraft. In other words, computers began to replace several systems that were electro-mechanical and also replaced some humans. That oversized room is called "The Firing Room." I had to work in that room to provide support for any problems that arose. Those of us who were Mission Support engineers had to undergo training in computer languages. The new computers were installed to provide the necessary services for the space shuttle launches. In 1977, we started to work on those systems in preparation for the launch of Columbia, the first shuttle to be launched into space. At that time, they were building five space shuttles—Columbia, Challenger, Discovery, Atlantis, and Endeavour.

I forgot to indicate that there was a space shuttle named the Enterprise. It was the first space shuttle prototype that came to the Kennedy Space Center. To be more precise, its first flight was on August 12, 1977, and I believe lasted a day. According to my notes, the last flight was on October 26 of the same year. Then that prototype went back to Rockwell International, which was the contractor that built all space shuttles. The prototype was redesigned so that it could fly in space as the others did. According to my memory, it was renamed Challenger. The first flight for the Columbia was April 12, 1981, and its final flight was on January 16, 2003. On that tragic day, Columbia broke up in the air. The first flight of Challenger was April 4, 1983, and the last flight, on January 28, 1986, exploded in the air. The first flight of Discovery was August 30, 1984, and its final flight was February 24, 2011. The first flight of Atlantis was October 3, 1985, and its final flight was on July 8, 2011. It was the last space shuttle that flew before closing the program. Fortunately, I was there from the beginning to the end of the program. The last three space shuttles were converted into museum pieces.

esa

european space agency
agence spatiale européenne
presents this
Team Achievement Award
to
JUAN P. RIVERA

In recognition of your valuable contribution to the First Spacelab Payload Project and with special appreciation for your professional dedication during the project culminating in a successful mission of Spacelab 1 on STS 9
28 November 1983 - 8 December 1983.

M. BIGNIER
Director Space Transportation System

E. QUISTGAARD
Director General

Figure 4: Award presented by ESA

NASA had already assigned the crew for the first space shuttle Columbia. The launching date was still up in the air. This spacecraft came in 1979 to Kennedy Space Center. All engineers were working tirelessly to make the first launch a success. We verified all systems, and, if something was wrong, we had to have an online replacement to assume the job at the push of a button. Finally came the day awaited by all. Managers decided the date of the first launch. It was April 12, 1981. I am not going to describe all the details, but we launched the Columbia, and the mission was a resounding success. It took four flights on different dates to prove that the space shuttle concept was a fact from the point of view of the required technology. As scientists, we took photographs and videos of all launches with an array of video cameras and numerous photo cameras. All these videos and photographs were analyzed, and the results were distributed among different teams of engineers and scientists. Launch pads A and B were used to support those space shuttle missions.

In July 1982, the Challenger arrived at the Space Center. Each of these spacecraft had to be inspected, and all the systems required for a successful launch had to be placed online. The third spacecraft called Discovery arrived at Kennedy Space Center on August 30, 1984. For me, this spacecraft was one of those whose systems had to perform perfectly. It had one of the most complicated missions to date. Imagine the space shuttle launching not one, but three communications satellites. For this flight, they had seven astronauts, the number considered adequate to take care of all systems for that mission. The human cargo, as well as the science supplies included in the flight manifesto, had a massive weight. However, despite a few hiccups, that mission was a resounding success.

In April 1985 came Atlantis, the last of these spacecraft. Atlantis was launched on October 3, 1985. It carried a payload that belonged to the Department of Defense. Not much was known by us of what was in the belly of that spacecraft because everything was controlled by that government agency. This trip was also successful.

I was fortunate to be part of the history of these missions—to see with my own eyes the latest technologies of the future before they entered the public marketplace. As an example, in early 1970, TV antennas (dish antennas) were enormous—usually 12 to 15 feet in diameter. The objective of any antenna is to capture a signal—in this case, a television signal. Television programs that originated on the opposite side of the planet (i.e., from Argentina) could be seen clearly in Orlando, Florida. Most of our neighbors had the immense antennas that cost up to $5,000. One day, my wife asked me to get one. I replied, "We launched a communication satellite in geosynchronous orbit. The receiving antenna needed is only 18 inches in diameter and costs about $100." Right now they are advertising those 18-inch antenna systems in my electronic magazines. To spend $5,000 on an antenna was financial folly. In less than a year, many people were using the popular 18-inch antennas. People who bought the jumbo 15-foot antennas lost their money.

Around 1982, NASA formed a new group of engineers called "Level Four." That group of engineers handled directly all the payloads that were launched using the Space Shuttle. I was fortunate to be part of this important group of engineers. It gave me the opportunity to travel the country from north to south and from east to west. My mission was to go to different space centers and universities that had experiments that would fly in the space shuttle. As I wore two hats—that of an engineer and that of a physicist—I combined all the necessary knowledge of the two professions in one person. My mission was to ensure that the experiments that were scheduled to fly on the shuttle complied with all requirements of NASA. Doing this project gave me the opportunity to rub elbows with the best scientists of this country and other countries. Years later, the work I was doing was distributed to other engineers. My future was going to change again.

I had the opportunity of being chosen by two groups of scientists on two separate occasions to join their technical team which spoke with the astronauts in the Mission Control Center in Houston, Texas. In that control room, there are many scientists working on a mission. A shuttle mission lasts several days and scientists are needed 24 hours a day for the duration of the mission. Therefore, we had to break into three shifts of eight hours. I often worked more than eight hours because my replacement was usually late. I enjoyed working at night, i.e., from 10 p.m. until 6 a.m. The main reason was that I was young and could take the punishment of the graveyard shift.

Figure 5: award presented by NASA

Figure 6: Award presented by SPACELAB

THE SPACELAB MISSION

One day, my boss, Mr. William, called me into his office to tell me that he was forming a new team to assist in a mission of European countries named the Spacelab. He wanted me to work on that mission because I was a scientist and also an engineer and I was the right person to work with European scientists. That meant that I was moving to another group within the group that I worked, i.e., a subgroup. I told my boss that I was ready to go where I was needed. This gave me a fantastic opportunity for a change of pace. The main reason is that when everything becomes routine, then it gets boring. I needed a job to challenge my intellect.

The Spacelab was actually a laboratory that included a series of experiments that flew on the space shuttle. Once the space shuttle reached its orbit, the bay doors opened so that all these experiments were exposed to space or the universe. The advantage of the Spacelab was that it could be modified according to the needs of scientists and the mission. Each flight could take different experiments to space. This laboratory had four components: (1) the pressurized module; (2) several pallets; (3) the tunnel module; and (4) a communication system. The main laboratory was in the payload bay of the shuttle in addition to the pallets.

The European Space Agency (ESA) was the owner of the laboratory. It was a contribution from different European countries including Canada. ESA gave to NASA as a gift the first laboratory, LM1. In return, NASA agreed to allow some European scientists to fly on the space shuttle. The following laboratory, LM2, was bought by NASA for its own use. My participation, as well as the rest of my teammates, was applauded by the administration of ESA and NASA. As evidence, I have included some plaques of that epic feat. Please refer to figures 6 and 7.

The pallets were used in 25 spaceflights. By mutual agreement of the parties, the team was decommissioned in 1998, and it was understood that the pallets could be used on the International Space Station (ISS). They built a smaller module, similar to that used in the Spacelab, to be mounted on the ISS. I understand that the laboratory LM1 is displayed at the airport of Bremen in Germany.

The pallet was a U-shaped configuration. The module consisted of a cylinder flying in the back (belly) of the shuttle, attached to the cabin of astronauts by a long tube. Both the cabin, the long tube, and the main module were pressurized in the same way as in an airliner. The laboratory diameter was about four meters and was composed of segments of approximately three meters long. The pallet was exposed to space and provided the base for placing the various instruments. At one time, the scientists built a telescope to observe the stars. We could say that was the forerunner of the Hubble telescope. The benefit of pallets is that they can be used as needed. They can use one, two, or three pallets, according to the requirements of the mission. I worked on the first few flights. The first one occurred in November 1983. We launched the second one in April 1985 and the third in July 1985. Thereafter, the same procedure was followed. I think the last pallet was launched in 2000.

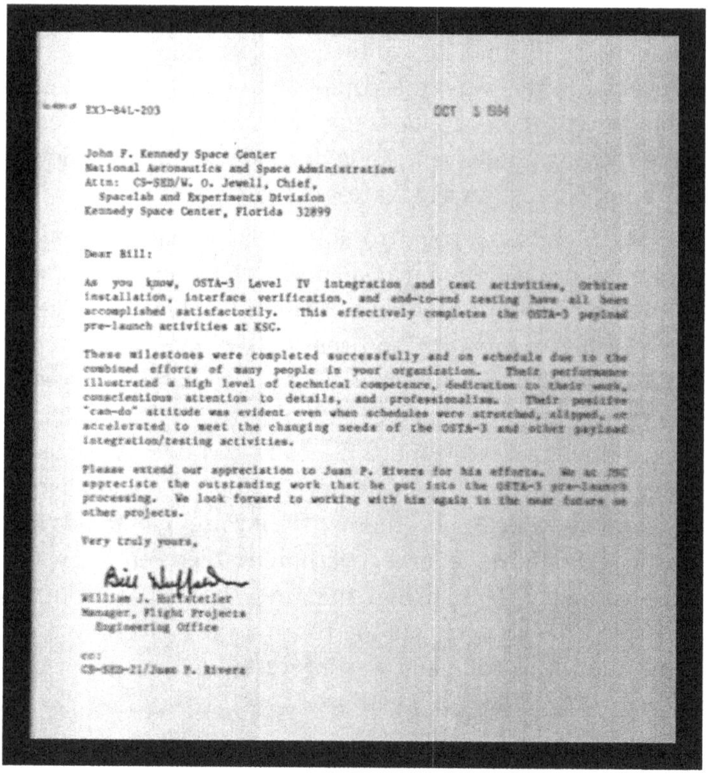

EX3-84L-203

OCT 5 1984

John F. Kennedy Space Center
National Aeronautics and Space Administration
Attn: CS-SED/W. O. Jewell, Chief,
Spacelab and Experiments Division
Kennedy Space Center, Florida 32899

Dear Bill:

As you know, OSTA-3 Level IV integration and test activities, Orbiter installation, interface verification, and end-to-end testing have all been accomplished satisfactorily. This effectively completes the OSTA-3 payload pre-launch activities at KSC.

These milestones were completed successfully and on schedule due to the combined efforts of many people in your organization. Their performance illustrated a high level of technical competence, dedication to their work, conscientious attention to details, and professionalism. Their positive "can-do" attitude was evident even when schedules were stretched, slipped, or accelerated to meet the changing needs of the OSTA-3 and other payload integration/testing activities.

Please extend our appreciation to Juan P. Rivera for his efforts. We at JSC appreciate the outstanding work that he put into the OSTA-3 pre-launch processing. We look forward to working with him again in the near future on other projects.

Very truly yours,

William J. Huffstetler
Manager, Flight Projects
Engineering Office

cc:
CS-SED-21/Juan P. Rivera

Figure 7: Johnson Space Center Award

MY LOCKHEED MARTIN YEARS

One time, an engineer friend of mine from Puerto Rico was looking for work and asked me to please help him at a job fair, where there would be representatives from various aerospace companies. The advantage of a job fair is that many companies send their recruiters to interview prospective employees. In this way, the company does not have to seek recruits to visit their facilities, thereby saving a lot of money. For the recruits, the advantage is that one can be interviewed by different companies in a matter of hours.

My friend was going to submit his resume to several representatives of different companies. It occurred to me to kill two birds with one stone, so I figured I would also take my resume to test the waters and see if they were interested in me. My friend was interviewed first and then the Martin Marietta official interviewed me. Before I left, I told the representative of Martin that I honestly was not interested in working for them because I was working for NASA and it would be extremely difficult for them to convince me to leave a fantastic job with the federal government to go venture with a private company. I told him that they had to make me a reasonable offer to persuade me to work for them. The agent told me that if his company was interested in me, they would call.

Within two weeks, my friend told me that he had been rejected. To my great surprise, I received a call from a representative of Martin Marietta (now Lockheed Martin) to invite me for a second interview. I agreed to a second interview, and, after negotiating a salary, I was quite satisfied. I decided to leave NASA and accept the position at Martin Marietta. I was already vested with NASA, but I had to wait until I reached the required age to obtain my pension. I spent five years at Martin Marietta working as an expert in physics and senior engineering. Readers who study engineering are suitable candidates to work with this exceptional company. Physicists with the right credentials can make a lot of money.

I became interested in the story of this famous company and would like to share its remarkable history with the reader. What follows is just a summary.

In 1926, Allan Loughead, Jack Northrop, and Kenneth Jay founded what is known as the Lockheed Aircraft Company. The headquarters was located in Burbank, CA. The company's main products were in the field of aircraft. Prior to World War II, they produced many aircraft. The Great Depression of the 1930s adversely affected the company, which went into a restructuring period and emerged as Lockheed Corporation under the leadership of Robert E. Gross around 1934. They continued building better aircraft because there were other legitimate players. The company built the famous Model 10 Electra Airplane which was a small twin-engine transport airplane. This was the plane that Amelia Earhart and her navigator Fred Noonan attempted to fly around the world in 1937. The company was extremely successful building airplanes during World War II.

Glenn L. Martin founded the Martin Company in 1912. He was an aviation pioneer. The company continued to grow under the leadership of Martin. The production of an aircraft was the first business initially. In order to remain competitive, these companies had to undergo some transformations. Around 1961, a new company was formed via another merger. The new company was named Martin Marietta. This is the company that I went to work for. The headquarters were in Baltimore, Maryland. The company's main products were missiles and rockets. This type of company is called an aerospace company. The Martin Titan program was necessary for NASA and still is a strong aerospace division of the company.

The main problem for an aerospace company (and, in fact, for any business) is what is known as a hostile takeover. The story of the large fish eating the smaller fish applies here too. Bendix Corporation tried to get Martin Marietta. In the end, Martin Marietta survived and Bendix was bought by Allied Corporation. When one is working for any of these companies, there is always the possibility of going to bed one night working for one company and waking up the next morning working for another one. Ownership changes hands often. For example, when I left Martin Marietta, I was already vested in the company, so I was expecting to receive a benefit from Martin Marietta, but keep reading to see what happened a few years after I left the company.

One day, my friend who was still working for Martin Marietta, sent me a message that Lockheed Corporation was trying to get the Martin Company. Around 1994, Lockheed Corporation merged with Martin Marietta to create a new company known as Lockheed Martin Corporation. The deal was finalized in March 15; 1995. This new business was actually a large company that controlled many segments of the industry. We are talking here of a company whose products include missiles, military aircraft, satellites and many others. Working for a solid company like this is certainly impressive. One has to be an extremely skillful person to survive in any of these companies because there are co-workers that envy your job. The only job security is your ability.

While working with this company, I had the opportunity to visit many places in the United States, traveling mainly to the West and Southwest. I also traveled to the northern states and the southern states. I have happy memories working for this company, but most of my work was classified. During these trips, I learned many useful things that could never be learned through books. If anyone is interested in a job in this field, more information about this and other aerospace companies can be found by searching on the Internet. Also, this company, along with others, frequently visits various universities across the country recruiting new engineers because their products are continually changing to meet the market demand.

INTERNATIONAL SPACE STATION

One day I was reading the classifieds and saw an ad from McDonnell Douglas Aerospace Company. They were looking for engineers to work in their Kennedy Space Center rocket division. I had all the qualifications for the job. In fact, I exceeded the minimum qualifications. So I quit my job with Martin Marietta in Orlando, Florida, and accepted the position with McDonnell Douglas Aerospace Division in Cape Canaveral, Florida. My responsibility was to apply my knowledge and working experience with NASA on a new project—the development of the International Space Station (ISS).

As I had worked on the development of the Space Shuttle Program, this knowledge would be pivotal in my new job. Once again, I saw myself beginning a new phase of my life. I had become a member of the team of engineers and scientists that would make the International Space Station a reality. I started working for this company in 1989.

The main purpose of a space station is to have a permanent laboratory in space. The Spacelab could not stay in orbit for longer than two weeks. This placed a limit on the scientific experiments. The ISS is a partnership of many countries—the United States, Canada, Russia, Japan, and the European Space Agency, hence the name International Space Station. Other countries may participate but must sign a contract with NASA. Italy is a good example of an entity that later joined the ISS. I believe that other countries are welcome to participate. The Russians had a space station called MIR that flew for long time, but the technology was outdated, and it had some other limitations. When I started working on this, there was no prototype; the goal was just an idea on paper. The name at the beginning was Freedom.

They had to create a new building to handle all the hardware that was to be launched into space. The astronauts had to work in space to assemble the different sections. It was estimated that the station had to have an orbit about 400 kilometers above the Earth's surface. The period of the orbit was calculated at about 90 minutes, which is the period of a satellite in synchronous orbit. The plane of the orbit was calculated at approximately 51 degrees to the plane of the celestial equator. It was designed so that at least two astronauts could stay there continually for up to six months. The biggest issue was the number of trips that had to be made to keep the station supplied with all that was needed, not only the experiments but the problem of food and human waste. Imagine living in the cabin of an aircraft flying at high altitudes for periods of up to six months. It certainly is not easy for two or more people to live in that kind of environment. Also, we calculated that the life of the space station exceeded the life of the space shuttles; therefore, we needed to have an alternate way to use Russian vehicles to replace the shuttles. Therefore, Russia's contribution was essential. It was estimated that the station would be inhabited by Russian, American and Canadian astronauts. With the right training, we could send other people—not necessarily just astronauts. We could call these individuals "Space Tourists."

While working for McDonnell Douglas, I started learning about the roots of the company. I was extremely familiar with how Douglas Aircraft Company products started the DC-3 series of airplanes that made history in civilian aircraft. Then I started digging more into the history of the company and the position the company has with their competitors.

Here is an overview of this great company. James Smith McDonnell and Donald Wills Douglas started the company in 1967. Before this date, each one had owned his own aircraft company—McDonnell Aircraft Company and Douglas Aircraft Company. Then sometime after the Second World War, they decided to combine the two companies and they named it McDonnell Douglas Aircraft Company. The two companies were officially merged in April 1967. The main reason for the merger was to be more competitive in the new field of rockets and missiles. Before the merger, McDonnell was essentially a defense contractor, and Douglas was mostly in the field of civil aircraft.

When I came to work for the company, we heard some rumors that Boeing was interested in acquiring the company. There were some problems that, at our level, we never understood extremely well. I believe that the ending of the cold war in 1990 created confusion in some companies. We learned that the curtailment in military procurements, plus the loss of contracts for two significant projects—the Advanced Tactical Fighter and the Joint Strike Fighter—severely hurt McDonnell Douglas. The Douglas MD-11 was notorious for having many problems. Many MD-11 orders were cancelled. Money was not coming in. At the same time, there was fierce competition with Boeing and Airbus. The idea of building bigger double-decker civilian airplanes was more in vogue in those days than in the past. We learned that Boeing acquired Rockwell North American which was the company that built the space shuttles. So then Boeing was getting bigger than the competition. Again, the large fish ate the little fish. This whole situation made me more concerned. I decided to get out of McDonnell Douglas and get my pension rights before the company collapsed. I left the company around 1996 with all pension rights. After I left, Boeing acquired McDonnell Douglas for billions of stock-swaps. McDonnell Douglas disappeared as a company, and Boeing was the surviving company. So now when people ask me, I say that I am a Boeing retired engineer.

Around 1993, I decided to start looking for another job because I knew that completing the space station would take many years. I worked five years on it, and honestly I was convinced it was a pretty complex design. It was calculated that it would take approximately 21 flights to construct the space station, using just the space shuttles. I was already getting on in age, and I felt I should use my resources in another exciting job before it was too late. I estimated that it would take more than five years for the space station to be completed. At that time, I was 55 years old, and if I waited longer, my job marketability would be zero. So I started looking for a new job. I was trying to find a job more attractive than the dreary space station project.

THE SCIENCE OF RADAR

One day, I received a call from a recruiter of Northrop Grumman Aerospace Company in Chicago, Illinois. They were inviting me to fly to Chicago for an interview. The company was going to pay all travel and hotel expenses. They offered me more money than what I was making, plus all travel costs and relocation expenses. The company would pay all moving expenses, including a 30-day stay in a hotel, and assistance to help me find a house or apartment in Chicago. My wife was happy. My youngest son was then five years old. I took care of all preparation details, and in two weeks, I was with my family in the famous city of Chicago. I felt as if I had won the lottery. Besides a generous salary, I had the convenience of living in one of the exclusive suburbs of Chicago. For those who have not been to Chicago, I will say that they are missing a lot. The best years of my life were spent as a resident of this city. Living in Chicago was the pinnacle of my career as a scientist and engineer. This company had a strong financial base and the best technical instruments imaginable. I was thrilled to have the opportunity to work with the most complex technical equipment, as well as with other highly qualified engineers and scientists. The company also offered exceptional fringe benefits to its employees. I had the opportunity to make several visits to Los Angeles, California, on company business. The headquarters of the company was located in Hawthorne, California.

At that facility, they also had modern equipment laboratories. Chicago has many attractions. For those who live there, and for those who wish to see this beautiful city, here is a short list of points of interest that I visited:

- Adler Planetarium
- The Art Institute of Chicago
- The Shedd Aquarium

- The Hancock Observatory
- The Field Museum
- The House of Frank Lloyd Wright
- The Children's Museum
- The Trolley Ride
- The Bus ride across the city
- The Sears Tower
- The Boat trip
- The Museum of Science and Industry
- The Navy Pier IMAX theater
- The Navy Pier Ferris Wheel

There are many other family attractions. There are also two ballparks, Wrigley Field (Chicago Cubs) and White Sox Stadium (Chicago White Sox). All sports were comprised of unusually talented teams. There are movie theaters and many more attractions. The biggest problem was having the time and money to see all the attractions of this incredible city.

I had the joy of working with Northrop Grumman Corporation (NGC), a company with a long history. It is made up of about 20 smaller companies (all more or less in the same market) that were once independent before NGC acquired them. NGC covered the entire spectrum of military battlefields, and their main customer was the U.S. Department of Defense. Other customers included many friendly countries. This company built or was involved in the construction of all types of aircraft, missiles, rockets, and other products. Many military and nonmilitary vehicles used all kinds of electronic instruments, whether on land, sea, or air, wherever the company plays a vital role. The company has been the cornerstone of success, from the famous flight of Charles Lindbergh to supporting the lunar landing mission. Future engineers and scientists with the appropriate qualifications can work for this magnificent company. I recommend it.

The company was founded by John Northrop, who was a virtuoso designer and pioneer in aeronautical apparatus in

Hawthorne, California. During World War II, the company built many types of fighter aircraft. Since then, the company has been one of the leaders in the manufacture of all types of aircraft and space vehicles. One of the most famous fighter aircraft is the B-2 bomber that, since 1989, has been protecting the free world. At one time, there was another independent company created by Leroy Grumman, along with some friends. They had the foresight to start a new aircraft engineering firm that became known as Grumman Aeronautical Engineering Company. During the Second World War, the company built many warplanes. In 1994, the people that once were the competitors of Northrop were acquired by them, and the combined company was called Northrop Grumman Corporation.

In 1938, Westinghouse's Radio Division was part of the leading company of Westinghouse. This division moved to Baltimore. They created the radar used in military equipment during World War II. One type of radar built by this section detected approaching Japanese bombers that attacked Pearl Harbor in 1941. This company built all kinds of electronic devices that were used during the war in different vehicles, whether on land, sea or air. After the War, the company continued building electronic devices for use in national defense. In 1996, the company was acquired by Northrop Grumman Electronic Division.

The word RADAR is an acronym for Radio Detection and Ranging. Let me try to explain how it works. A transmitter sends a radio signal which travels towards a target. The signal bounces off the target and returns to a receiver. We measure the time that it takes the signal to travel back and forth. The apparatus measures the time of flight of the signal, and using basic physics equations, the machine calculates the speed of the target. We use the Doppler Effect principles to determine if the target is moving away or approaching the device. This extraordinarily versatile machine can detect the presence of more than one target, determine whether it is approaching or receding, and determine if it is an enemy or a friend. It can tell if it is an aircraft and what kind of aircraft it is. It can decipher if it is a friendly signal or an enemy signal. Being a physicist is an advantage because I learned the technology extremely fast. The science of radar is of primary importance not only for military applications, but also for other applications.

Radar is used to detect geological fractures or any other failure in the soil. Other types can discover with absolute precision the presence of metals, water, minerals, and other materials below the ground. They are also used in agriculture for tracking the relative humidity of the crop and the effectiveness of irrigation systems. They have many uses for mapping and working in trees due to pests or diseases in trees. One of the most prominent uses is tracking the seabed by those seeking Spanish ships full of gold, coins and other valuable minerals. Many Spanish ships have been located, and treasure hunters have been able to recover merchandise from Spanish ships that carried treasures from Mexico, Peru, and other South American countries. Radar is also used to detect illegal fishing. In summary, radar has many scientific applications.

As one may have seen in this book, all kinds of engineering are based on fundamental principles of physics. Technology is a tool that scientists and engineers use to solve complex problems. My advice to readers who want to make money is to study engineering. Being a competent engineer is essential to have complete control of the principles of physics. To study physics, one must have a strong mathematical background; if one is not an expert mathematician, then one cannot be a competent engineer.

When I began to study physics and mathematics, electronic calculators and computers had not yet been invented. We used slide rules. Years later, electro-mechanical calculators followed by electronic calculators came into the market. After that came the massive IBM 360 computer, followed by mini-computers and then the desktop computer. It is necessary to remember that technology changes at a fast pace and one has to be up to date on all advances in technology. I was born in 1938 before the Second World War, and since then, I have seen all these technological developments. Thanks to the Divine Creator, I had the excellent fortune to witness the progress of all that technology and be part of its development. When I went to work for this great company in Chicago, I was 56 years old. I had the opportunity to learn the science of radar and other electronic devices that were a challenge for my intelligence. With a solid educational foundation and strength of will to overcome all obstacles, one can win the game.

Thankfully, I was able to work as an engineer and scientist for Northrop Grumman. By the year 2000, I reached my retirement age. My time had come to say goodbye. The company made me a reasonable offer. It was hard to say goodbye, not only to the company, but also to the great city of Chicago. But I had to be realistic. Although I could have stayed for many more years, I decided to retire with a decent pension while still happily enjoying perfect health. Looking back, I would say that working with this company was a treat for me because it was my last post as a scientist and an engineer. After I left, the company continued to grow, and I think it will continue to acquire other companies of considerable importance. The company helped me with my moving expenses back to Orlando, Florida. For the first time, moving was bearable, because the only thing I had to do was watch to see that the movers did not break any of my belongings. At that time, my older children had become independent of me, and I just had my youngest son and my wife, who traveled first to Orlando. I stayed for six months in Chicago working with the company. Once my wife had bought a house in Orlando, I took a plane and joined the family.

MY ENCOUNTER WITH A UFO

They say that the best is left for last, but this is not actually the case. In order to understand my encounters with the extraterrestrials, the reader must first have the necessary objectivity to be able to accept the veracity of these encounters. The writer is a scientist, an engineer, and a university professor who has worked for many years in various disciplines of human knowledge. I have worked in my profession for more than 50 years. I have a strong religious foundation. As a scientist, I use daily the scientific method. I do not believe in a story unless it passes a rigorous scientific analysis using the principles of logic. The author is not a deceiver, but a truthful person who thinks a lot before exposing any idea. What I am about to relate happened many years ago. I've had to keep it a secret because disclosure of the facts would have hurt me professionally in the various aerospace companies that I've worked for. I was also afraid that many people might label me a fool or a person who has lost his mind. Now that I'm close to the end of my life, I have decided not to carry my secret to the grave. At my age, I have no fear of anything. I only respect the Divine Creator who has protected me throughout my whole life.

To understand this, it is vital to understand the meaning of certain words that I will use. A UFO is an "unidentified flying object." That is, many people (including me) have seen a real flying object that is unfamiliar. It is not a plane or a helicopter or a comet or a kite. Extraterrestrial beings are intelligent beings that come from somewhere else in the universe. They are also known as aliens. Ufologists are those who believe in UFOs. What I am about to describe is not a dream. It is a fact. It is as real as the moon or the sun.

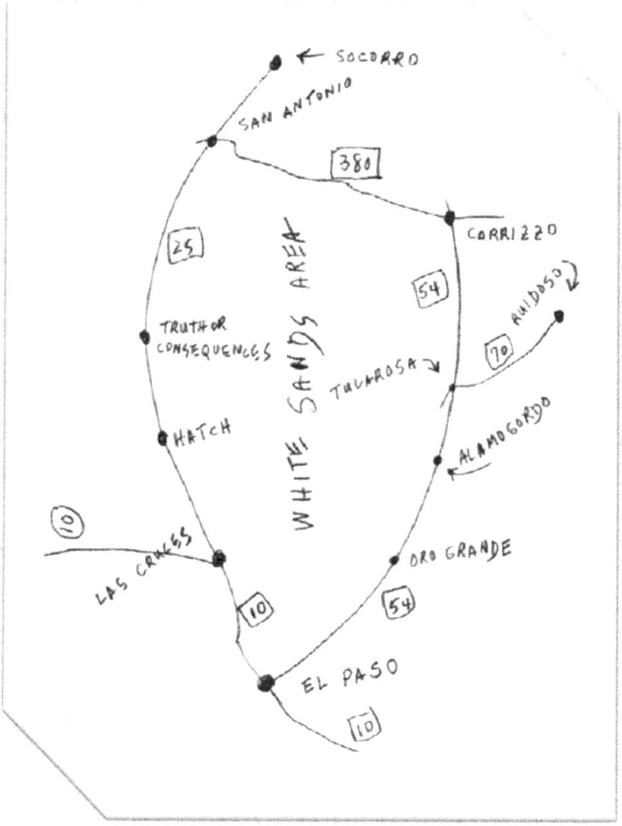

Figure 8: White Sand map

Many years ago when I was young, I was working as an scientist in White Sands, New Mexico. This is a remote area in the state of New Mexico. To get an idea of its location, please refer to the map shown in Figure 8. The closest villages at that time were Alamogordo and Ruidoso. The bigger city of Las Cruces also had an entrance to this testing area. At that time, I had to travel after work on U.S. Hwy 70, a road that was singularly desolate. I had to go through Alamogordo and Tularosa because I lived in Ruidoso. At other times, I had to use another route that took me to another entrance and passed through Carrizozo, where one can see the famous statue known as "Trinity Site." The official name of this area of

New Mexico is White Sands Missile Range (WSMR). It is extremely well known that this area is dedicated to the testing of military hardware. The public can reach a park that is available on certain days. There are now a few monuments and even a museum that tourists can visit. The park also has a gift shop where one can buy souvenirs. In this place, there is a park of missiles and rockets. I have seen many flying missiles and rockets through my life in many places. People interested in seeing these rockets and missiles usually go through an entrance which is near the city of Las Cruces. There is also a gate which is near the city of El Paso, Texas. I was familiar with the whole area in my youth when it was smaller and more remote. I have not been back to this area in many years. I believe that there are sectors that are not available to the public.

A long time ago, one could travel those roads and not see a car for a long while. At night, the setting was more dangerous. I was afraid to travel those roads at night. Going to work in the morning, I could see more cars heading to the White Sands area. The beauty of the desert and the mountains was outstanding. There is a series of mountains which is known as "The Organ Mountains." The name suggests to me the male genitals, but I could be wrong.

One night, I was on my way to Ruidoso and noticed that a flying object was following me. It looked like a spaceship, i.e., a UFO. I thought I was seeing a mirage. As a good scientist, I made a turnaround to prove to myself that it was not a mirage or an optical illusion. I saw that the ship was in the form of a disk. Being familiar with the colors and the pattern of lights that aircraft use in this country, I found the colors of the disk to be quite unusual. Aircraft lights are normally green, red and yellow. But the lights on the spaceship were green, red and blue. Suddenly I entered into an emotional state of fear for my life. My hands were trembling. I could hear my heart pumping faster. I hit the gas pedal to the floor, hitting speeds of 100 miles per hour, but the ship was still following me. The main problem was that it was around 10 p.m. at night. Suddenly I saw a white light in the distance. It appeared to be a neon light at the end of the road. Gradually, as I got closer, I saw that it was a gas station. Luckily, I saw a New Mexico police patrol parked at the entrance. When I looked in my rear view mirror, I noticed that the ship had disappeared. I decided to stop and talk to the policeman.

The police asked, "What can I do for you?" I told him all that had happened to me. The cop looked at me, shining his powerful flashlight on my face. "How many beers did you drink tonight?" he asked. "None," I replied. Then I showed my company identification card. I told him that I was a scientist working in the WMSR and I was heading home after work. The policeman looked at me this time more pleasantly and indicated a willingness to help. He said in a friendly tone, "Get in your car and I'm going to give you an escort to your house. I promise that nobody is going to kidnap or injure you." The police car followed me all the way until I got to my apartment in Ruidoso, New Mexico. I told the whole story to my wife because she was wondering why I came home so late. Then I went to bed. I could not sleep because I was still shaken.

The next day, I first talked to the security guard at the gate where I had to enter WSMR. Security guards at that time were men of Mexican descent, and both spoke excellent English and Spanish. More importantly, they knew who went in and out of that place. They always checked if there were people working before they shut the gate. I asked him what time they closed last night. He pointed to a sign by the gate that showed opening and closing times. He said, "Look man, I think that after you left, no one was left behind because we closed this gate and then went home." Well, I still had doubts in my mind, so I decided to see the base commander. I got an appointment for the next morning at 9:00 a.m. I went to talk to the military officer who knew everything that was going on in that place. He asked, "How can I help?" As I told him my account, I noticed that his face was turning red. Suddenly he stopped me and said, "Look, please never say anything to anyone because I promise that you will lose your job. You sound crazy. Do not disclose this story to anyone." Well, I left the office of the commander, and then I realized that my job was in danger, so I did what he told me and I didn't say anything to anyone. As my grandmother used to say: "No flies can enter a shut mouth." As a scientist, I had my doubts myself because I believed that the long hours of work were affecting me. I continued doing my job for several weeks and tried exceptionally hard to forget the incident.

One night I was driving back to my apartment. I left the gate at work and said goodbye to the security guard and continued driving towards my house. This time I was on Route 380 to Carrizozo. From there I took Route 54 toward the town of Tularosa. The time was again around 10:00 p.m. It was intensely dark, and the road was singularly desolate. All of a sudden, I was aware of red, green and blue lights shining on my chest. Then I noticed that in a cow pasture there were many cows resting. In this area of New Mexico, there were many landowners whose main business was selling livestock. Then I noticed that there was a cow lying on the ground. I saw that an aircraft, shaped in the form of a circular disk, had landed nearby. The ship had three legs equidistant from one another that gave support to it, similar to the NASA Lunar Lander module. I heard a terrible strange noise. I didn't know if it was coming from the aircraft or the little men, but, certainly it was not a familiar earthly sound. I turned off my car lights. Silently, I sneaked up to get a better look.

I saw three little men dressed in green suits. Their blue heads were shaped like an upside down pear—wide at the top, less wide around the mouth. I estimated their height at four feet. I looked at their eyes and noticed that their pupils were yellow and their irises were red. Their eye sockets were very dark. (Please refer to the front of this book.) Their tongues were forked like those of a snake, but instead of two points, their tongues ended in three points. I saw that they were drawing blood (possibly from the jugular vein of a cow) using their tongues like hypodermic needles. They had two hands and two feet, but what struck me was that they had a third eye on their foreheads. Looking carefully, I noticed that it was not a third eye, but rather like a light bulb that they turned on and off to see better. The light was green, reminding me of fireflies that I saw as a child in Puerto Rico. One of them noticed my presence and started running towards me. I got so scared that I ran like a gazelle down that dark highway. I ran and ran because I thought he was coming after me, but I got tired of running and felt like my heart was coming out of my mouth. When I finally stopped to rest, I saw that nobody was chasing me anymore. I looked up at the sky and saw no stars and no moon, and I could not see my hands. Panic turned to relief when I realized that the three little men and their UFO had disappeared. I got into my car and drove towards my apartment. That night I did not tell my wife anything. I never said anything to my co-workers or to the base commander's office people. I just ignored it for a while.

Years later, I met my former NASA co-worker George. He was the only person that I could trust. He and I used to share stories about the unknown. He was extremely familiar with other UFO sightings in other parts of the world. He was an Ufologist. I felt extremely at ease and told him what had happened to me at White Sands, New Mexico. It was then that he showed me a medallion. He asked me if any of the men in green uniforms was wearing a medallion centered on their chest. I was stunned. I told him that the three little men with blue heads had medallions hanging from their necks. The medallions shone even when the moonlight was not present. I reiterated to George that the blue-headed men had what looked like a third eye on their foreheads and that they could turn it on and off at will like a firefly. I made a drawing for George. Then he went home.

When I was working at the laboratories at Kennedy Space Center, George used to bring me objects for chemical analysis. One day he brought me the medallion. I had almost forgotten that incident. I worked in a NASA lab, and I had access to the best scientific instruments to investigate these strange objects. With the consent of my supervisor, I submitted George's medallion to a chemical analysis using a spectrograph. The analytical results showed that the medallion metal was not of this planet. The atomic weight and other chemical properties were not listed in the periodic table. He had bought the medallion from a Mayan Indian on one of his trips to Central America. I took a picture of it. I still have the picture.

Another day, George showed me a picture of another medallion and asked me if I could decode the symbols. I showed the picture to my boss who was an expert in the field, and he asked me if I had brought that picture from Mexico. I told him that George had asked me to explain the symbols. My boss told me that he had seen similar symbols in the area of Chitchen-Itza, and the symbols seemed to be Mayan in origin. He had a theory that the Mayan leaders had left this planet. There is another theory that the Inca Indians had gone from this planet to an unknown location in the universe. He stressed to me that those were only theories. The symbols on the medallion were similar to those of the Mayan Indians. Of course, I could not reveal my encounter with the aliens to my boss because I would have lost my job on the spot.

MORE EXTRATERRESTIAL TOPICS

As I said earlier, when I worked for NASA, I had the opportunity to work with some astronauts. Some of these astronauts then became my friends, and in the course of more intimate communication with them, I try to discuss UFOs and the existence of extraterrestrial beings. As they understood my technical terms, I noticed that whenever I brought up the issue of UFOs, they changed the subject. It seems that they could not discuss certain topics which could reach the public domain. There was a rumor among those who worked at the Kennedy Space Center that astronauts have seen UFOs. One of them was interviewed by newspapers, magazines and other media. I once read, if I remember correctly, that astronaut Gordon Cooper had talked about UFOs to the news media. I have not been able to substantiate such a claim. The rumors at the space center always sounded true. One might think that the astronauts would be the first witnesses of these UFOs, but all these things have to go through a sieve and everything stays in limbo. In the mainstream government media there seems to be censorship on issues of this nature. People who believe in UFOs protest that these topics are prohibited by the government.

According to some sources, astronaut Edgar Mitchell, who was a crewman on Apollo 14, said he had seen UFOs. However, NASA officially denied it. The blue-headed men that I saw had enormous eyes and heads shaped like a pear. There is no doubt in my mind that I was looking at the real thing. I must point out that these extraterrestrials did not appear to have any hostility towards me. As I am also an astronomer, I must say that, from the point of view of astronomy, there must be other intelligent beings like us in other galaxies. It has been proven by astronomers that there are other planetary systems like ours in other galaxies. There are even planets in those galaxies that have characteristics similar to our planet earth.

We earthlings belong to the galaxy that is commonly called the Milky Way. Astronomers using the Hubble telescope's immense power, as well as other instruments, have clearly discovered that there are other galaxies. It is assumed that in each galaxy there is a system of planets. There must be planets similar to earth's living conditions. My opinion is that the existence of extraterrestrial beings is a fact. They must have a certain level of knowledge and their bodies would have perhaps a different morphology. They created a civilization similar or superior to ours. I think that the governments of this planet earth must maintain the exploration of outer space to seek other civilizations that might exist.

When I first visited the Arecibo Radio Telescope Observatory in Puerto Rico, its primary purpose in 1960 was just to find evidence that extraterrestrial beings existed. Unfortunately, at that time, I was too young to understand the importance of this phenomenon. Although I had the opportunity to stay there, my adventurous spirit prevailed and would not let me.

The incident that took place in Roswell, New Mexico, is another example of aliens that came to earth. The government has always kept this episode under a veil of secrecy. Undoubtedly, there occurred "something" that has never been clear and that was censored by the mainstream government media since its inception. At that time, the technology was quite inferior to the one we have now. What really crashed in that area of New Mexico?

It is also said that when Armstrong got off the lunar module Eagle, he found that he was not the first to set foot on the moon. He saw strange objects not belonging to the moon as if aliens had already been there, and Armstrong was only one of many who had visited the moon. So if this is true, where did these objects originate? Did they come from Russia? Did they come from another galaxy? NASA officially declared that all of these claims were false. In the same way, I can neither confirm nor deny what happened on the moon. However, they say that Edgar Mitchell has been one of the few astronauts who have publicly acknowledged the existence of UFOs. I can confirm what I saw, but I cannot endorse or reject what others saw. My boss told me that people who believe in aliens should be in an asylum. Of course, I did not challenge his comments but kept my silence because my job was at stake.

When I was working at White Sands, New Mexico, I was aware that area 51 existed because some of my friends worked there at one time or another. The Air Force and other official agencies that knew about the project altogether denied its existence. It was considered taboo to discuss the subject.

Just recently the Central Intelligence Agency confirmed the existence of Area 51. But, instead of clarifying the situation people think that they added more uncertainty to the subject. Now they say that the secrecy of the spy planes and high tech spy aircraft sparked American imaginations. In other words, what people call UFOs are actually high-tech spy aircraft. Well, if this is true then why they have spent millions of dollars in the Arecibo Observatory that its main goal has always been looking for the existence of extraterrestrials or some kind of intelligence in outer space. What I saw was not a U-2 or a drone or a spy plane. None of those spy planes is in the form of a circular disk. None of the spy plane can sit still in the middle of the air for a long period of time. They cannot make a right angle turns. None has legs like the lunar Lander module for landing purposes.

The people that built the pyramid in Egypt, in Central America, in Bolivia, India and the other part of the world use tools that were more advanced than the tools of people living on earth at the time that they were built. Let me ask the following questions. Why the extraterrestrials abandoned these magnificent sites. Where are they? Are they coming back? Why the bible talks about the ascension of Christ to the heavens? Or his return to earth? From the Astronomy point of view where is heaven or hell? Are this just symbolism or are they real sites that exist somewhere in the universe? These and other mysteries need explanations. I have to clarify that many religions believe in a place called **heaven** and a place called **hell.** Catholics also believe in the existence of the **Purgatory** and another place called the **limbo**. I respect all religions and this book is not about religion. I am just asking questions.

Also, people in Central and South America have reported seeing UFO's shaped like disks, yet those countries have few if any involvement in world politics to become a threat to the United States.

Many people that believe, that alien are connected with natural disasters. It is also known that technology can be used to influence the weather. For example; Nikola Tesla was trying to create a device which could send an electromagnetic signal that will bounce in the ionosphere and be reflected back to the earth. This is the principle for short wave radio transmission and reception. Some people believe that Tesla was also trying to create a device that could affect the weather conditions. Some examples are; thunder, lightning, tornadoes and even hurricanes.

If an engineer or scientist can possess such technologies, then it will be logical to think that extraterrestrial intelligence possess also such technology. Some people believe that aliens are responsible for tsunami, hurricanes, tornadoes, lightening, powerful storms and other natural disaster. Such information has been published in the media. Those interested can do their own research about this subject.

My NASA friend George, told me that he read a book, which according to the author a group of humans were transformed into a race of grey-skinned beings, stunted and with oversized heads. According to George, he read another book that the author talks about the "Selenite" (natives of the moon) that have grey skin, oversized heads and large black eyes and a wasp-sting.

George also told me that one of his friends read another book in which the author describes a race of extraterrestrial whose heads were big and bald, with square foreheads and extremely small noses, mouth and weak chins. They have big eyes, dark and gleaming with a sharp gaze. They wore soft grey uniforms and limbs similar to humans. So in the media Ufology media these extraterrestrials are called THE GREYS.

In 1965, the Betty and Barney Hill abduction brought back "The Greys" to international attention. They claimed that they were abducted by these extraterrestrial people who took them to a saucer shaped spaceship. Their description of these extraterrestrial people has common traits to the so called GREYS. Later on it was claimed that these extraterrestrial people originated in the Zeta Reticulli star-system which is exceptionally well known to astronomers. I believe that later on the famous 1977 movie "Close Encounter of the Third Kind" the Greys were involved in the plot.

During the early 1980, many ufologists linked the GREYS to the alleged crash-landing of a flying saucer in Roswell, New Mexico. There has been plenty of documentation related to the Roswell flying saucer crash-landing. UFOs sighting has been documented also by millions of people in probably every country of the world. This means that I am not the only person that has reported an encounter with extraterrestrials. The main difference is that the extraterrestrial that I saw were from another Galaxy in the universe because they were wearing green uniform. Their skin was blue (looked purple under the night sky). We must conclude that based on that we cannot call them the Greys.

On one of my trips I went to Puerto Rico because they were talking about the appearance of EL CHUPACABRAS. Then I met Rosa, which told me that around 1993, Fox television aired a series called "The X-Files" which made a combination of the search for proof of extraterrestrials with the so called UFO CONSPIRACY THEORY. So I will say that there has been much media coverage of the extraterrestrial theories which mostly are connected to the Greys and some mention of EL CHUPACABRAS. There have been also claims of documented film or video of an autopsy of a Grey extraterrestrial. For those readers that are not familiar to the chupacabra, let me clarify that issue.

My friend Rosa, from Puerto Rico told me that, around 1995, the CHUPACABRA was some kind of an animal that soaked (chupa) blood from the goats (cabras). So somebody coined the expression of the CHUPACABRA. They said that eight sheep were discovered dead, and each one of them showed three puncture wounds in the chest area that probably some animal soaked their blood. Since the aliens that I saw, have their tongue ending in three points, I thought that there was some connection.

I immediately make arrangement to fly to Puerto Rico, but, by the time I got there the sheep was gone. After that date, there were other people that reported seeing the creature in the town of **Canovanas**. Many animals were killed in the same fashion. All of them showed the distinctive three puncture wounds in their bodies. In 1975, several animals were also found dead in the same fashion in the town of **Moca**. There have been more reports of animals killed in the same fashion by draining their blood. Each animal show the distinctive three puncture wounds in their bodies. Reports of similar animal being killed the same way have been reported in many states, in the USA and Mexico. More reports followed in other parts of Central and South America. Even in Russia they reported many animals being killed in the same fashion.

The vision of the cow that I saw in the middle of the night being bled by the extraterrestrial makes me think that probably what people called CHUPACABRA may have a connection with the extraterrestrial that I saw that night in New Mexico.

There have been many claims that connect NASA with aliens as I have discussed previously. There has been a lot of discussion with the alleged Alien base that was discovered in lunar photos taken by the NASA's Apollo 11 mission in 1969. There have been many claims that the alien base in the moon is as real as the moon or the sun.

Many people insist that NASA or other responsible government agencies are trying to cover up the proof for the existences of aliens. The search for extraterrestrial Intelligence was one of the first objectives of the Arecibo Ionospheric Observatory as I have mentioned earlier. It has been reported that many scientists, astronomers and other technically minded people are continually working actively in the pursuit of this goal.

The "Search for Extraterrestrial Institute" SETI is a non-profit organization whose prime mission is to explore, understand and explain the origin, nature and existence of life in other parts of the universe. They used both radio and optical telescope to search for signals being sent by extraterrestrial intelligence to earth. So, the Arecibo Radio Observatory is really a big ear 1,000 feet in diameter that is listening for the source and nature of these signals. The Carl Sagan Center is pursuing the Study of Life in the Universe. This include; (1) the discovery of extra solar planets, (2) potential for life on the planet Mars, and (3) life in other celestial bodies in our Solar System. Among other things, SETI is working with NASA on exploration missions such as the Kepler Planetary Mission and the SOFIA Mission.

There have been reports of massive alien spaceships heading toward earth. Other people claim that these are undoubtedly meteorites or other types of extraterrestrial bodies. There is even a United Nation Panel on Extraterrestrials. Some people speculate that there might be some indications that aliens from the Planet Zeeba are in the process of invading the planet Earth.

There has been reports that NASA's Planetary Science Division has collected information which when studied could shed more light to the question of extraterrestrial intelligence. Nobody knows at this time what would happen when the so called aliens from the Planet Zeeba confront humans living in planet earth. There are people that believed that aliens are already among us. That they have infiltrated our institutions and that someday they will decide to attack us with unknown results. There are many people that believe that extraterrestrial may assist us in the solutions to diseases, poverty and hunger. On the other hand, there are people that believe that there is a significant possibility that aliens will wipeout our civilization.

There are ufologists that claim that there is a group of researchers that are sending messages into space. They are using the Arecibo Ionospheric Observatory to connect with aliens in a peaceful manner. There is a belief that the rapid expansion of our civilization may be extinguished by aliens to conserve a balance.

My NASA friend George told me at one time that that the Honorable Paul Hellyer, which is a former Canadian Minister of Defense, publicly admitted that he believed that our planet earth has been visited by aliens for years. According to George, "The Huffington Post of Canada" published that Hellyer testimony stated that "aliens live among us and that at least two of them worked for the U.S. government." The allegation continues that Hellyer also declared that "four extraterrestrial from Zeta Reticully, the Pleiades, Orion, Andromeda and Altair star constellations have been visiting us for many years with different agendas." Hellyer publicly affirmed that "alien technology obtained at the crash-landing of the UFO in Roswell, New Mexico, could be used to develop substitute to newer fuels in order to break our dependence of fossil fuels as our prime energy source."

I have read that the **Pumapunku site in Bolivia** is another proof of extraterrestrial civilization. The precision technology used by this civilization is out of this world. They landed in what is known today as Bolivia in South America. The high precision of the small holes drilled in stone is astounding. The heavy block of stones with a set of blind holes drilled in complex shape is a prominent example of the high technology used by these extraterrestrials. The Pumapunku or **Puma Puncu** is quite a large temple complex or group of monuments that are part of the Tiwanaku Site near Tiwanaku, Bolivia.

The area near **Tiwanaku** is where the Incas believed that the world was created. In the Aymara language, Pumapunku means "The Door of the Puma." The Pumapunku is a grandiose complex set of monuments or buildings which consist of a western court, a central esplanade, a terraced platform mound that is faced with a megalithic stone and a walled eastern court. These heavy stones have been cut with some kind of sophisticated high precision tool unknown to the civilization of that era. The "Plataforma Lirica" is located in the eastern edge. This platform consists of voluminous heavy stones in perfect shape that was cut with high precision with some unknown tool. The terrace is paved with enormous stone blocks that also were cut with probably the same high precision tool. Based on the density of sandstone it has been estimated that the stone slab weighs approximately 131 tons! The crucial question is; how were these stones moved from one place to another? What machine they used to place these stones in their precise location? It is hard to believe that the Incas possessed the diesel engines machinery to move a massive stone block from point A to point B. So the conclusion is that this site was built by extraterrestrials.

Anthropologist has excavated the area and came to the conclusion that the site was built during three different time frames. So the question is this site built by the same civilization or different civilizations? To complicated matters we now have to face another crucial question. During its peak, Pumapunku was adorned with polished metal plaques. This civilization used brightly colored ceramic. They use ornaments made of some form of fabric. It is believed that the priests and their elites dressed with elaborate exotic jewelry. The question is where this civilization went to? Are they coming back? Why they didn't destroyed these sophisticated site.

This area of Pumapunku in Bolivia, has been studied using the best technology like; ground-penetrating radar, magnetometers, induced-electrical conductivity and magnetic susceptibility instrumentation. Using all of the data collected by scientific methods has revealed that the area between

Pumapunku and Kalasasaya is rich in numerous structures made by an unknown civilization. The structures include wall foundations, water conduits, pool-like features, beautiful terrace, residential compounds, and widespread use of gravel pavements streets. So who built this site? Many believed that the Incas did not have the tools to build such an elaborate site. The majority of ufologist thinks that the site was built by an extraterrestrial civilization.

Archeologists have concluded that the materials used to build this site were transported from a highly inclined quarry near Lake Titicaca. This quarry is roughly about 10 kilometers away. The question that arises is what device they used to move such heavy stone up a steep incline?

Archeologists also point out that other materials came from quarries within the Copacabana Peninsula which is about 90 kilometers away from and across Lake Titicaca. So another question is what device or vehicle was used to transport these heavy stones through such a large distance?

Some archeologists believe that the transport of this large and heavy stone was accomplished by using large labor force of Inca Indians. There are other theories that have been proposed. Two of them seem to have more scientific explanation. One is that they used llama skin ropes, the other, is that they use of either ramps or inclined planes. We can speculate that only extraterrestrial have the knowledge and capability of accomplishing such an enormous task.

Archeologists also explain that the walls of Pumapunku, where built like a jigsaw puzzle. Each stone was precisely cut to interlock with the surrounding stones. The fit is astonishingly highly precise! **They did not use any kind of mortar!** The forming load-bearing joints fit in such a way that has lasted for many centuries. They have witnessed the inclemency of weather and other natural disasters.

Since I am an engineer, I understand the gravitational forces that come into play when building anything above ground. One technique used in engineering involves cutting the top of the lower stone at a certain angle, and put another stone on top of it which was cut at the same supplementary angle. In order to do this type of joint, one must have a good knowledge of solid geometry and physics.

Those that work in masonry know that these cuts must be precise that not even a razor blade will fit between the stones. When I was studying the basics of masonry, I remember that much of this skill is characterized by accurately cutting rectilinear blocks with such uniformity that they could be interchanged for one another while maintaining a level surface and even joints. That is why that masonry is not an easy task and needs a lot of precision and expertise. The archeologists point out that the stone block was precisely cut. They indicate that prefabrication and mass production was crucial in this type of building. These technologies were so advanced that is hard to believe that the Incas were the ones that built this site with the technology of that era. We need to admire what I will call the Tiwanaku engineers that constructed this site without using mortar and on top of that they constructed; (1) a functional irrigation system, (2) the hydraulic mechanisms and (3) a waterproof sewage lines system.

There are many people that have studied the complexity of the Pumapunku site, and then by comparison studied the Pyramids of Egypt, or the Pyramids in Central America. In the same way that we ask ourselves who built Pumapunku, then we ask who built the pyramids in Egypt or the many pyramids in Central America.

Many ufologists believe that the best explanation is that aliens also built these many pyramids that although they are very far away from each other, they have many similarities.

My friend Rosa told me that a Cairo University archeologist said that UFO Aliens may have helped build the Pyramids of Giza. Dr. Ala Shaheen, head of the Cairo University Archeologist Department allegedly told an audience in 2010 that "there might be truth to the theory that aliens helped the ancient Egyptians build the oldest of the pyramids, that is, the

Pyramids of Giza." It has been said that Mr. Marek Novak, a delegate from Poland riposted to the question; did the pyramids might contain alien technology or even a UFO within its structure? Rosa told me that he said that "he cannot confirm or deny this, but, there is something inside these pyramids that is, **not of this world."** Some people in the audiences were left shocked, however, Dr. Shaheen had refused to comment further or elaborate on his UFO and alien related statements.

There is also an allegation that, the "Secret KGB UFO Files Documentary", deals with the fact that the Russian had already discovered the tomb of an Alien Humanoid in Egypt and an that "something" is beneath the pyramid which might be "not of this world." According to Rosa, "The Secret KGB UFO Files" documentary supports that the head of the Cairo University Archeology Department, Dr. Ala Shaheen claim is true.

My NASA friend George told me that the idea that aliens exist has been hidden by governments for many years, yet, the sixth man to walk the moon, Dr. Mitchell, claimed during a radio interview that "he was aware of many UFO visits to Earth during his career as a NASA astronaut, but, each one was covered up." Dr. Michell was an astronaut during the Apollo 14 mission to the moon. He describes aliens as "little people who look strange to us." He continued saying that "real life extraterrestrials were similar to the traditional image of small frame, large eyes and head." He stated that "our technology is not nearly as sophisticated as theirs." He added that "had they been hostile, we would be been gone by now."

Dr. Mitchell, along with Apollo 14 commander Alan Shepard, holds the record of the longest ever moon walk, at 9 hours and 17 minutes. According to Rosa, after that memorable mission he said; "I happen to have been privileged enough to be in on the fact that we've been visited on this planet, and the UFO phenomena is real." He continue saying; "It's been well covered up by all our governments for the last 60 years or so, but, slowly it's leaked out and some of us have been privileged to have been briefed on some of it." Dr. Mitchell, who holds a Doctor of Science degree in Aeronautics and Astronautics, claimed that; "Roswell was real and similar alien visits continue to be investigated." George told me also that In a radio interview with host Nick Margerrison he said; "This is really starting to open up. I think we're headed for real disclosure, and some serious organizations are moving in that direction." They say that Mr. Margerrison said; "I thought I'd stumbled on some sort of astronaut humor, but he was serious that aliens are definitely out there, and there's no debating of it."

Officially NASA was quickly to respond by saying; "NASA does not track UFOs. NASA is not involved in any sort of cover up about alien life on this planet or anywhere in the universe." They added that "Dr. Mitchell is a great American, but, we do not share his opinions on this issue."

If you look in the night sky, you will not miss three bright stars that shine with a brilliance that is very peculiar. The Orion constellation is a pattern of stars that is unique from any other constellation. In the belt of Orion are three stars that have great significance to ancient civilizations. The pattern of these three stars is commemorated in many monuments through the ancient world. These monuments can be seen from Egypt to Mexico. Our ancestors were intrigued by these three stars. Now the question that we must ask is why was Orion important to the ancient civilization from Egypt to Chitchen-Itza? Why are these pyramids built and oriented to those three stars?

In central Mexico, the ancient city of Teotihuacan is aligned astronomically to those three stars! All pyramids are oriented 15 to 25 degrees east of north, and the front wall of the Pyramid of the Sun is exactly perpendicular to a point in the horizon where the sun sets on the sun equinoxes. The other ceremonial buildings in Teotihuacan are at right angles to the Pyramid of the Sun. The Avenue of the Dead is pointing to the setting of the Pleiades.

The Dog Star Sirius which is sacred to the ancient Egyptians suggests a link between the great pyramids of Egypt and Mexico. There is no question to me that I have seen both pyramids that these pyramids were built by the same people. In fact, when I look to some Egyptians, and some Mexican Indians there is no question in my mind that both of them are related, yet, they live a great distance from each other. How the Egyptians or Mexican Indians traveled from one place to another? What type of vehicle they use?

As an astronomer that I have been, I have observed many times to the constellation of Orion. If you look to the night sky from November to February, Orion is in the southwest of the night sky if you are in the northern hemisphere. If you are in the southern hemisphere, you will see them in the northwestern sky. This constellation is a partly a circumpolar constellation that can be viewed most of the time, (depending in what latitude the observer is located.) For those readers that are interested in viewing this constellation, the right ascension is 5 hours and its declination is 5 degrees. The name of the three stars are as follows; Alnilam, Mintaka, and Alnitak. These three stars are commonly known also as the Orion's' belt. They are the most visible stars of this constellation.

Betelgeuse is the second brightest star in Orion, which is in the right shoulder of the hunter. On the left shoulder is the star Bellatrix. The most famous is the Orion Nebula which is really a formation of dust, hydrogen; helium and other ionized gases rather than a star. This is the so called "star" in Orion's sword, which hangs off of Orion's belt. The Horse Nebula is also nearby. There are other stars, which are called Hatsya, Melissa, Saiph and Rigel.

I should point out the constellation Cygnus is also very famous. This constellation is also known as the Swan. Ancient astronomers including Egyptians and Mayans were attached to this constellation. Many people called it also the Northern Cross. The Aztec, Mayans and Incas watched very carefully the night sky. They believed that they came from that part of the sky and that they are heading after death to that part of the sky.

In 2006 and thereafter the Hubble Space Telescope has dedicated many hours to the study of this part of the night sky. When one looks at the astronomy literature, one will see splendid images of the Orion Nebula. The detail photo shows more than 3,000 stars in that region in various stages of life, and it has given researchers new information of these celestial bodies. We acquired vast knowledge of other galaxies and planetary systems. The galaxy that we live on is called the Milky Way. Our planetary system is part of the Milky Way galaxy, but, there are other galaxies. There is also planetary system in each galaxy similar to ours. So we believe that there are other planets out there that have similar characteristics to the planet earth. Therefore there exists intelligent life in those planets. We must conclude that we are not alone in this universe. There is much other civilization out there.

The Orion constellation is the most known star formation in the night sky and is mentioned in the ancient cultures many times. To the Greek ancient civilization Orion was part of their mythology. When the hero was killed he was placed among the stars for eternity. Poseidon is the father of Orion in Greek mythology.

If one studies the Egyptian mythology, one will understand their religious belief better. The gods descended from the Orion Belt and Sirius to earth. They came in the form of humans as Osiris and Isis. They instigated the human race. Osiris and Isis started the Egyptian civilization and, in the long run, the whole human race originated from it. According to ancient Egyptians Orion was linked with the creation of the world. These beliefs are shared with Incas, Aztecs and Mayan civilization. So there is another connection between Egypt and Mexico.

It is believed that Osiris will return from Orion one day, and not only in Egypt have we found tales of "Gods" promising to return we also know that Christians are expecting Jesus

Christ to return. According to Christianity Jesus Christ ascended to heaven and will return to earth. So there might be another connection with Christianity and the belief of extraterrestrials.

The concept that the Pyramids in Egypt and Mexico are aligned with Orion's belt is really amazing. Now my question is why the Great Pyramid has air shafts that point to Orion? Why the pyramids in Mexico are also in precise alignment with Orion?

Napta Playa is a site that lies in the eastern region of the Sahara Desert. It was discovered by a team of archeologist around 1974. The researchers found stones scattered in a site that seems to be part of an ancient civilization that probably existed in 6400 to 3400 B.C. In other words, this civilization existed before the Egyptians.

It is believed that NABTA PLAYA is really not a settlement. But, what has intrigued archeologist is that this seems to be like a "mini Stonehenge" out in the middle of the Saharan desert. There is a circular stone structure that has an intricate alignment to the stars of the Orion constellation. Whoever were these people there is no doubt that they had a good knowledge of Physics, Astronomy and Mathematics. They built these structures in perfect alignment with the Orion constellation.

My NASA friend George told me that according to Robert Bauval, an engineer and Thomas Brophy an astrophysicist that have studied very carefully this intrigued monument for several years, they suggested that the stone circle is like a diagram that was used to view the stars. The diagram aligns with the belt stars Orion at the summer solstice instead of the equinox. The mysterious smaller circle is like a diagram that seems to explain how the sky moves. It seems that the small Nabta Playa observatory is some sort of calendar circle. Using radiocarbon technology scientists have estimated that the site was built approximately 7000 years ago. Using the collected data Brophy estimated that the three center stones lined up precisely with the three brightest stars in the constellation of Orion around 4900 B.C.

Another question that is boiling in my mind and it forces me to ask; if there is Aliens in Outer Space, then is there Aliens in Heaven? In the medieval epoch, there was the belief that the earth was the center of the universe. So if you study astronomy you have to study the cosmos according to Ptolemy and his vision of the universe. In order to explain the motion of the planets Ptolemy proposes a vision of the universe were the earth was at the center and the planets, the sun and the moon revolves around the earth. This system was known as the Geocentric Cosmology. Humanity was considered as the primary manifestation of god. Well this seems very logical because we observe that the sun, the moon moves across the sky from east to west. In this vision of the universe the earth is stationary and everything moves around the earth. The problem was explaining the retrograde motion of the planets. In other words planets are moving from east to west then all of a sudden they change direction and move from west to east. Then later on they resume their motion from east to west. To explain this, Ptolemy has to invent what he called the epicycles. So his vision of the cosmos was able to explain the motion of the planets using the epicycles theory. The rest is history.

Then the question came, are we along in this planet? So the Catholic Church was forced to add other creatures like the angels and archangels, heaven, and hell. Other vision of the cosmos took shape. Then came the complexity added by Christopher Columbus and his discovery of the new world. So now we have to include in the cosmos, the new world Indians. So the poor Indians needed to be added to the vision of Christianity. So things continue to complicate as years go by.

The discovery of extraterrestrial life would present Christianity with a big obstacle. So theology needs to either be modified or reject the idea of the universe as we know it today. With the addition of the space shuttle, the International Space Station, the different space probes sent to the infinity of our universe then the Christian vision of the universe needs to be modified again. Many churches are forced to deny the existence of extraterrestrials. So the question is; will the church add another form of theology to accommodate the existence of extraterrestrial? Or will they take the path of least resistance and completely ignore the whole idea of extraterrestrials? Will aliens be aware of the Divine Creator as we know it? Do aliens believe in another form of religion? Are alien atheists? What about the Big Bang Theory? Will God have placed life intelligence elsewhere in the Universe?

One day I was discussing this topic with Felipe, another friend from Mexico, who also was intrigued by these questions. According to him, the reverend Jose Gabriel Funes, head of the Vatican Observatory and who was the scientific advisor to Pope Benedict XVI was interviewed by a newspaper. In that interview he expressed; "How can we exclude that life has developed elsewhere? Just as there is a multiplicity of creatures on earth, there exist other beings, even intelligent, created by God. The Alien is my brother." We don't know if this was his opinion or the official church opinion. So apparently we have more questions than answer regarding the relationship between religion and extraterrestrials.

My NASA friend George and I had many setbacks because of what I have narrated here. I lost contact with my other friends Rosa and Felipe. Unfortunately, George died and I don't know what happen to the medallion, but I still have a photo of the medallion. On another occasion, I will speak of other events that occurred. Both he and I did a lot of research on these topics. Many groups_have invited me to discuss these topics, and I always refused to do so because my job was at stake. Now I am able to discuss these and other topics and probably give lectures, provided the time and my health will allow it.

BECOMING A PROFESSOR AGAIN

Since 2000, I returned to academics. The reason was and is simple. I have a Doctorate in physics, and I am also an engineer; I would not have any problem in finding a job as a professor of physics or mathematics. I am used to working in the academic field. I have taught several courses in different fields of knowledge such as physics, mathematics, astronomy, programming and computer applications, engineering courses, and many others. My hobbies are photography, videos, electronics, musicologist, and Ufologist. I helped as a volunteer, and continue doing so, for Hispanic organizations in Central Florida.

My message to my friends and the reader is to never look for excuses. If one wants to succeed in life, there are no obstacles to prevent it. Will is Power. My message to the parents of these young people is that they need them. There are no excuses not to help your sons or daughters who need your help now. I was an extremely poor boy born in a neighborhood surrounded by violence, crime, poverty, and slowly I got out of the hole and got a decent education that helped me overcome many obstacles. Some of my students and friends have asked me what the greatest achievement of my life is. My answer to them is always unexpected. The greatest accomplishment was moving from Rio Piedras across the Atlantic to Miami on a barge. From there, I continued a long journey of several days to reach the capital of Mexico. I think I've been the only fearless person to make that trip. It is one starting a journey from the known to the unknown. (When Christopher Columbus left Spain, he made a long journey never knowing what to expect.) To achieve a goal, one must have a lot of willpower and belief in oneself to make that journey. The funny thing is that to do that one does not need to go to school. Not only did I make that long trip, but I was forced to undertake another trip from Mexico to Albuquerque,

New Mexico. On top of that, I made another trip from Albuquerque, New Mexico, to Huntsville, Alabama. Last, but not least, was the one that I made from Huntsville, Alabama to Orlando, Florida. I think I've been the only person who has made all those heroic journeys. I thank God that for all those trips I had a "Guardian Angel" that protected me. I hope this story is an inspiration to many of the readers of this book. I ask anyone to please take this book to someone who might need a little help.

Apparently, this book ends here. However, this book is the beginning of my next book in which I will provide more details of a third encounter I had in New Mexico with extraterrestrials and the mystery of the medallion. Also, I will discuss why there is so much censorship on anything that has to do with extraterrestrial beings. I will talk about the secrets of nature and the world. I will discuss new discoveries in astronomy and NASA's plans to explore other worlds in space.

Made in the USA
Monee, IL
04 November 2026

67385475R00134